Is American Science in Decline?

Is American Science in Decline?

Yu Xie and Alexandra A. Killewald

Harvard University Press

Cambridge, Massachusetts, and London, England | 2012

Library of Congress Cataloging-in-Publication Data

Xie, Yu, 1959–
    Is American science in decline? / Yu Xie and Alexandra A. Killewald.
        p.   cm.
    Includes bibliographical references and index.
    ISBN 978-0-674-05242-0 (alk. paper)
    1. Science—United States.   2. Scientists—United States.   I. Killewald,
Alexandra A., 1983–   II. Title.
    Q180.U5X54   2012
    509.73—dc23   2011042030

To our spouses, working scientists:
Yijun Gu and
Phillip Killewald

Contents

Acknowledgments

The origin of this book can be traced to Yu Xie's dissertation, "The Process of Becoming a Scientist," which he completed at the University of Wisconsin–Madison in 1989. The dissertation was about social determinants that affect a person's likelihood of becoming a scientist. Since his graduation, Yu Xie has been teaching at the University of Michigan, where he has maintained a research program on social stratification, statistical methods, social demography, and Chinese studies. Although in 2003 he published, with Kimberlee Shauman, a coauthored book titled *Women in Science: Career Processes and Outcomes* (Harvard University Press), he had left his dissertation on the process of becoming a scientist unpublished. Some professional friends liked his dissertation, however, and made inquiries about it over the years.

Eventually, Yu Xie decided that he would publish the dissertation as a book if he could find a capable graduate student as a collaborator to update the data analysis. In 2006, Alexandra (Sasha) Killewald, then Sasha Achen, joined the University of Michigan as a doctoral student in public policy and sociology. Sasha was a perfect match, as she shared interests with Yu Xie in social stratification, science, and, most important, empirical understanding of social institutions and human beings. The end result of our collaboration is this book.

Our original plan to revise Yu Xie's dissertation did not work. The world has changed, and so have our intellectual interests. We persevered, however, and ended up with an entirely new book. While this book is still concerned with social factors that affect who becomes a scientist, its primary focus is on the current policy debate concerning the state of American science. While many articles, reports, and books have recently been published on this debate, in researching this book, we were surprised to discover that they had not fully utilized vast amounts of data currently available on related issues. Too often, authors seemed to hold preexisting views or jump too quickly into assuming a policy position on the debate over whether American science is in trouble. For this book project, we mined these existing data as thoroughly as we could and also collected new data of our own. As much as possible, we tried not to let our own personal views influence our analysis and to allow the data to speak for themselves.

We are indebted to many individuals who have influenced or helped us. First and foremost, we would like to thank the members of Yu Xie's dissertation committee: Charles Camic, Warren Hagstrom, Robert Hauser, Charles Manski, and Robert Mare. Next, we are deeply grateful to many of our associates and assistants for their important contributions to the book. Cindy Glovinsky reviewed several drafts of our manuscript and helped greatly in shaping the final version. Her mark is on almost every page of the book. Qing Lai conducted most of the analyses in Chapters 7 and 8 and assembled the tables and figures presented in this book and in the supplementary online material. Daniel Hom and Chris Near were mainly responsible for the analyses in Chapter 4. Qing Lai, Chris Near, and Chunni Zhang provided feedback on an early version of the manuscript. Eileen Divringi helped with references and contributed ideas to Chapter 2. Mark Murphy provided statistical analyses for Chapters 6 and 7. Crosby Modrowski collected the historical data on media coverage of science reported in Chapter 5. Glenn Firebaugh at Pennsylvania State University was gracious in sharing with us the Penn World Tables, which were used in Chapter 2. Rhonda Moats's clerical support was also valuable. A Changjiang Scholar research grant at Peking University funded the valuable research assistance of Jingwei Hu, as well as the additional proofreading support of Guoying Huang, Sha Ni, and Yongai Jin. Miranda Brown provided moral support and consultation on historical materials for the book.

Funding for this project was provided by the Alfred P. Sloan Foundation. Additional funding came from the University of Michigan's Distinguished University Professorship research fund, the Survey Research Center, and the Population Studies Center, which is funded by a center grant from the National Institute of Child Health and Human Development (R24HD041028). Various cores of the Population Studies Center provided valuable support to our project: the Computing Core staffed by Ricardo R. Rodriguiz, David Sasaki, and Mark Sandstrom; the Data Services Core headed by Lisa Neidert; and the Information Services Core headed by Yan Fu.

Two anonymous reviewers provided thorough and very useful comments that helped us improve the manuscript. Michael Aronson at Harvard University Press, always a supporter of the project, was instrumental in making this book a reality. We would also like to thank John Donohue and all those involved in the production of this book.

Finally and most importantly, we are deeply indebted to the members of our families for their unreserved support. We dedicate this book to our spouses, who are both working scientists, the very group that this book is about.

Supplementary material for the book is available online at http://www.yuxie.com.

Introduction

> The dominant position of the United States depended substantially on our own strong commitment to science and technology and on the comparative weakness of much of the rest of the world. But the age of relatively unchallenged US leadership is ending.
>
> —NATIONAL ACADEMY OF SCIENCES, NATIONAL ACADEMY OF ENGINEERING, AND INSTITUTE OF MEDICINE, 2007

Science is of tremendous importance today. It has long been valued not only for its own sake as knowledge of the natural world but also for its direct contributions toward improving countless aspects of human life, such as the availability of food, water, housing, and material goods, health, education, communication, transportation, and security from natural disasters, epidemics, and human warfare. Closely related to science is technology, which is sometimes considered synonymous with applied science. Although it is not easy to pin down precisely what the economic impact of technology is in any given society, there is little controversy over the general sentiment that "technological change—improvement in the instructions for mixing together raw materials—lies at the heart of economic growth."[1]

The crucial role of science in an economy is widely recognized and acknowledged.[2] Indeed, the increasing importance of science is a major reason that today's society is characterized as a "postindustrial society" and today's economy as a "knowledge economy."[3] While social scientists may debate how to accurately assess the economic returns to national investment in science and technology, among policymakers and the public there is widespread agreement regarding the importance of science and science-based technology as an economic engine.[4] Sixty-nine percent of Americans believe that scientific research is "very important" to the U.S. economy,[5] and most introductory macroeconomic textbooks follow the work of Robert Solow, who presents technological change as the primary mechanism by which economic growth can be sustained.[6]

Science and technology are the results of human activity. Although practiced by individual scientists, they can have a large impact on a society. Thus, the importance of scientists in today's world is not a matter of contention. To sustain economic growth in the modern age, any society, including the United States, must possess a large and talented scientific labor force. The recent

1

global economic crisis that began in 2008 has intensified public awareness of the critical importance of technology, as policymakers, desperate for solutions, increasingly turn toward scientists and engineers for solutions to economic problems. "America's economy is in crisis," said former U.S. Senator Edward Kaufman on the Senate floor in February of 2009. "We can either drown under the weight of the problem, or we can ride the wave of opportunity that it offers. To do that, we must put science, engineering and innovation back in their rightful place in our economy."[7]

One reason why technological change has had a dramatic effect on past economic growth—and may have an even more dramatic effect in the future—is a unique feature of scientific knowledge in general: such knowledge can, with little effort, be shared at virtually no additional cost to either those who produced the knowledge or those who adopt it.[8] Although most published materials and inventions are legally protected, scientists have a long tradition of sharing scientific ideas freely with one another through journal articles, books, conference presentations, unpublished papers, and, more recently, over the Internet.[9] Good ideas are shared, bought, copied, emulated, leapfrogged, and sometimes even stolen.

Thus, in satisfying their own thirst for knowledge and pursuing scientific activities, scientists as a group exert an enormous positive impact on their societies.[10] For this reason, many policymakers believe that it is both legitimate and economically rational for governments to subsidize scientists' work. While scientists may benefit directly from such subsidies, spillover effects to the larger society may well justify the cost. Hence, support of science is different from other types of government support that mainly help the beneficiaries. As a common expression goes, "Give a man a fish, and you have fed him for today. Teach a man to fish, and you have fed him for a lifetime." Science and technology are like knowledge of fishing: once distributed, they can feed many people for life. In providing economic well-being to their people, scientists in many countries, including those in the developing world, do exactly that.

While science is an occupation for those individuals who practice it, it is not just any occupation. Although scientists are usually rewarded for their achievements, the true beneficiary of the work is not just individual researchers but the whole of humanity, which receives countless direct and indirect benefits from science through complex and varied pathways. For this reason, all industrialized countries are concerned with the condition of science and its practitioners—scientists—within their borders. In this book, we will examine scientists within the contemporary U.S. context, asking who our scientists are, who is most likely to become a scientist, and how American scientists as a group are faring.

The Debate: Is American Science in Decline?

The twentieth century was "The American Century."[11] By the end of the century, America had become the world's only superpower, leading the world in many areas. Much of America's economic prosperity and world power have been derived from America's leadership in science and technology throughout most of the twentieth century.[12] Defying a Japanese historian's bold prediction in 1962 that "the scientific prosperity of [the] U.S.A., begun in 1920, *will* end in 2000,"[13] American science is still going strong today.

Comprising only 5 percent of the world's total population, the United States can claim responsibility for one-third to two-thirds of the world's scientific activities and accomplishments on most measurable indicators. For example, in the world today, the United States accounts for the following:[14]

(1) 40 percent of total research and development spending;
(2) 38 percent of patented new technology inventions by the industrialized nations of the Organization for Economic Cooperation and Development (OECD);
(3) 45 percent of the world's Nobel Prize winners in physics, chemistry, and physiology or medicine through year 2009;[15]
(4) 35, 49, and 63 percent, respectively, of the world's scientific publications, citations, and highly cited publications;
(5) 85 percent of the world's top 20 universities and 54 percent of the world's top 100 universities.[16]

Individually, these statistics are measured with error and may not necessarily be accurate indicators of America's contribution to science. Taken as a whole, however, they are persuasive in presenting the United States as the continuing, unchallenged world leader in science, technology, and innovation. No other country has reached levels on these indicators anywhere near those of the United States,[17] as American scientists themselves seem to realize. In a survey conducted in 2009 by the Pew Research Center and the American Association for the Advancement of Science, 49 percent of the American scientists surveyed rated America's scientific achievements as "best in the world" and 45 percent as above average compared with those of other industrialized countries.[18] In addition, 88 percent of the same group believed that U.S. achievements in their own specialties were either the best in the world or above average.[19]

While most experts agree that U.S. science is dominant at present, many researchers—as well as policymakers and the general public—have become less confident regarding its future prospects. Questions have been raised not only as to whether the United States can sustain its leadership in science in the future but also as to whether the country actually overestimates its current strength in science relative to that of other countries.[20] One recurrent concern is over the potential shortage of scientists in the United States, a debate that has surfaced from the end of World War II to the present, beginning in the 1950s in response to the Soviets' launch of the satellite Sputnik into orbit and reappearing in the 1980s as national concern with mathematics and science education grew. In 1990, former University of California president and National Science Foundation director Richard Atkinson, in a well-publicized article in *Science* magazine, predicted " 'significant shortfalls' of scientists in the near future" and declared this "a national crisis."[21] In another issue of *Science* the same year, Philip Abelson, famous for his joint discovery of neptunium with Edwin McMillan, linked excellence in science to the overall well-being of a nation. "In the future global competition," he wrote, "a country not tops in chemistry is destined to be second-rate or worse."[22] Note that these two quotes embody two distinct but related claims: that an impending shortfall of American scientists exists and that national well-being depends on scientific dominance. For simplicity, we call such sentiments the "alarmist view." This view has been widely shared, as well as criticized, by observers and commentators.[23]

The alarmist view was recently embodied in a highly profiled 2007 report issued by the National Academy of Sciences (NAS), the National Academy of Engineering, and the Institute of Medicine, titled *Rising above the Gathering Storm: Energizing and Employing America for a Brighter Economic Future*. The report was drafted by the Committee on Prospering in the Global Economy of the 21st Century, organized in 2005, in response to a bipartisan congressional request to assess the current state of American science, identify urgent challenges, and recommend steps to be taken to make sure that the United States retains its scientific leadership. The report, released within weeks after the committee's first meeting, emphasized the economic importance of science and engineering. While it maintained that the United States is still the "undisputed leader in the performance of basic and applied research" as well as in applying scientific advances in improving the state of the economy, the authors expressed concern that the present state of affairs may not continue, due both to science being on the rise in other countries and to inadequate national investment by the United States in educating scientists and supporting them in their endeavors. Failing to keep up scientifically may have devastating effects not only on the economy, the authors said, but also on public health, the environment, national security, and other aspects of American life. The report's recommendations focused on improving K–12 and higher education, investing in long-term basic scientific research, and creating environments for research and innovation that will attract and retain the best and brightest students in the world.[24]

This report received a great deal of attention from policymakers, spawning over two dozen bills in Congress within a year of its release. Several indicators, in both education and the labor force, do seem to validate its concerns. In education, an increasingly large share of U.S. science degrees, especially at the PhD level, go to persons who were born abroad (as we discuss in Chapter 7). American high school students' performances on certain mathematics tests have been interpreted as mediocre by international standards in recent decades, suggesting that students may be ill prepared for advanced scientific training.[25] In the labor force, the rising share of immigrants among practicing scientists and engineers indicates that America's dependence on foreign-born and foreign-trained scientists has dramatically increased.[26] Furthermore, science and engineering have been criticized for low rates of participation among women and minorities, which can be viewed as a problem of both inequality and underutilization of talent.[27] A recent report under the auspices of the National Academy of Sciences urged efforts to increase the participation of nonwhite, non-Asian U.S. citizens in science as a way to increase both the diversity of perspectives represented in science and the supply of native-born American scientists in case the recruitment of immigrant scientists to the United States becomes more difficult in the future.[28]

However, the concerns expressed by the authors of the original NAS report and others did not go unchallenged. Scholars at RAND responded to the report by holding a major conference, followed first by the publication of the proceedings,[29] and, the following year, by a more unified report,[30] which, while acknowledging some areas of concern, such as the relatively low test scores in mathematics and science of U.S. K–12 students, challenged the NAS report's claim that science in the United States is experiencing a "creeping crisis" as well as its "clarion call" for drastic action to save America from the devastating effects of scientific and technological decline. Dire warnings issued in the 1980s and 1990s, the RAND report pointed out, were not followed by the anticipated catastrophes, and thus it is important to evaluate the data carefully. The RAND report specifically challenged the notion that globalization of science and technology and increased capabilities of other countries are inherently harmful to the United States, arguing, instead, that these trends may be beneficial.[31]

Along with the authors of the RAND report, other scholars maintained that the state of American science and/or science education was less worrisome than the NAS authors had claimed. In a working paper published in 2007 by the Urban Institute, Lowell and Salzman attacked the NAS report by using data from the National Assessment of Educational Progress, the math portion of the Scholastic Aptitude Test, the Trends in International Mathematics and Science Study, and the Program for International Study Assessment, to paint an overall picture of U.S. math and science education as healthy and steady.[32]

Moreover, economic studies have found little direct evidence of a market shortage of scientists practicing in the United States.[33] We will provide our own analysis on the issue later in this book, showing that, in recent decades, growth in the scientific labor force has outpaced growth in the general labor force, and scientists' earnings have not grown relative to those in other professions, as we would expect in the case of a shortage. In fact, erosion in the economic rewards to science as compared with those to other professions, due both to falling relative wages and increasing career launching time for scientists, has led some scholars to assert that, while America may face a scientific crisis, it is a problem of low demand rather than of low supply.[34] An article that appeared in *Miller-McCune* in 2010 challenged the NAS report by pointing to the current insufficiency of jobs for qualified scientists, citing a 2005 report by the National Research Council titled *Bridges to Independence: Fostering the Independence of New Researchers in Biological Research*, which described a "crisis of expectation" among young scientists frustrated by the lack of career opportunities. The 2010 *Miller-McCune* article also cited a finding that only a third of the Americans earning degrees in science and engineering each year are able to find work in their fields. In contrast to the alarmist shortage view, this perspective instead suggests an oversupply of young scientists compared with the demand for scientists in the workforce.[35] We will discuss this aspect of the debate more fully in Chapter 8.

Despite these criticisms, the main message of the NAS report that America needs more and better science has resonated well among business leaders and politicians. A report released by the Education for Innovation Initiative in 2008 described business leaders as being concerned about the lack of policy action to maintain scientific leadership in the United States.[36] In his address to the National Academy of Sciences in April 2009, President Barack Obama declared unequivocally that the nation had fallen behind in science, at least in terms of its investments:

> A half century ago, this nation made a commitment to lead the world in scientific and technological innovation; to invest in education, in research, in engineering; to set a goal of reaching space and engaging every citizen in that historic mission. That was the high water mark of America's investment in research and development. And since then our investments have steadily declined as a share of our national income. As a result, other countries are now beginning to pull ahead in the pursuit of this generation's great discoveries.[37]

The president's remarks heralded a large infusion of federal funds into scientific research as part of the American Recovery and Reinvestment Act of 2009. The scientific community welcomed President Obama's commitment to science and technology.[38]

Meanwhile, the criticism of the original NAS report did not persuade its authors to reverse their earlier claims. In fact, the authors reaffirmed their alarmist view in a follow-up report in 2010, declaring that "our nation's outlook has worsened" and describing the "gathering storm" as now "rapidly approaching Category 5."[39] This report discusses changes since 2005 (when the original report was being drafted) that are likely to have a negative impact on the competitive status of American science in the future. These include the continued growth and greater investment in science in other countries; the global economic downturn, leaving less money available in the federal budget to address the challenges of American science; and the financial crises of America's universities. The new report concludes with a chapter on "The Ingredients of Innovation" in which, based on data gathered since the 2007 report, it maintains that American science is now in even greater peril than it was in before.[40]

In evaluating the claims of the alarmists versus those of their critics, it is relevant to recognize that any changes to national science policy would have differential impacts on different interest groups. For example, an increase of scientists would be beneficial to the business sector, which could then enjoy the productivity of scientists at lowered labor cost but would intensify competition for employment among individual scientists and thus lead to the worsening of their labor force outcomes. We note that the committee that produced the 2007 NAS report included not only university presidents, Nobel Prize winners, and former presidential appointees but also a few former corporate CEOs. Their critics, in contrast, have often been academic economists who are keenly aware of a perennial shortage of tenure-track positions relative to the number of aspiring young scientists.

Whether or not America is headed for a catastrophic shortage of scientists, it would be difficult to deny at least some degree of vulnerability in this respect today. American science and technology could be damaged if the flow of immigrant scientists should stop or dramatically decline or if falling wages for scientists should begin to deter American-born youth with talent in science from pursuing scientific careers. Given the potential risk to our economy and national security, it is prudent to ask whether there is a current shortage of scientists in America today or whether America is in danger of experiencing a shortage of scientists in the near future. The alarmist view—that there is now or soon will be a shortage of scientists in America—is essentially a proposition about social trends. That is, the current period is viewed as a disappointing departure from a past era in which American science enjoyed unquestioned glory. At least according to some observers and commentators, Americans' lack of interest in science represents a cultural change from the era immediately following World War II, when American society as a whole became fascinated with science following the Soviet Union's launching of

Sputnik.[41] Hence, to assess the alarmist view, it is important that we examine empirical data that pertain to changes concerning scientists and potential scientists in the last few decades in America. Fortunately, a wealth of data have been collected by government agencies and research organizations that can be mined and brought to bear on the research questions of whether—and why—only small numbers of American youths pursue scientific careers. In this book, we present our analyses and interpretations of these data.

In the final analysis, how optimistic or pessimistic one is regarding the future of American science may depend on what criteria one chooses for evaluating its many different aspects. In judging the health of science as a profession, researchers have looked at the performance of K–12 students on mathematics and science tests, the percentage of freshmen planning to major in science or engineering, graduate enrollments and doctorates awarded in science and engineering, rankings of research institutions, number of employed scientists, number of publications, number of patents applied for or awarded, number of papers listed in the Science Citation Index, number of Nobel Prize winners, percentage of the gross domestic product spent on research and development, public and media attitudes toward science, percentage of immigrant scientists, and a host of other factors. In this book, our primary focus will be on the number of employed scientists in the United States and on those factors that influence the process of becoming a scientist, such as the rewards scientists receive for their work, the individual characteristics of those recruited into science, cultural attitudes toward science, and the quality and availability of scientific education. Of course, we are under no illusion that a thorough understanding of these factors will necessarily enable us to resolve, once and for all, the debate over whether American science will continue to thrive or decline. As we shall see, American science may be robust by some indicators but face a potential decline by others. Thus, in this book, we hope to move beyond black-and-white generalizations about the health or decline of American science toward a more nuanced, but more accurate, assessment based on the full complexity of its various aspects.

Definitions of Science and Scientists

To address the question of whether a sufficient number of youths in contemporary America are attracted to scientific careers, we need first to define both "science" and "scientist." By "science" we refer to a special kind of knowledge that is generated and verified through the use of standardized methods and that involves systematic experimentation and reasoning, often with the help of technology and/or mathematics.[42]

Two definitions of "scientist" are common in the literature: (1) the behavioral (occupation-based or demand-based) definition; and (2) the credential (education-based or supply-based) definition.[43] In practice, the occupation-based definition specifies that the incumbents of scientific occupations are scientists. The education-based definition considers individuals with or working toward science degrees as scientists or potential scientists.

In this book, we use both definitions. When we discuss trends in employed scientists, we make use of the occupation-based definition. When, however, we examine scientific pipeline issues—questions of how individuals progress through the various stages of scientific education and from education to scientific careers—we study science education and thus invoke the education-based definition. Operationally, we include engineers, computer scientists, and medical scientists as part of the definition but exclude medical doctors, computer programmers, technicians, and social scientists. Our choice to exclude the latter groups is based on the interest in the literature in the potential shortage of native-born scientific talent in the hard sciences.[44] Whenever possible and meaningful, we conducted our statistical analyses of scientists in the labor force and potential scientists in the educational pipeline with a focus on four major fields: biological science, physical science, mathematical science, and engineering.[45] Note that computer science is considered part of mathematical science for most of the analyses, although we sometimes analyze this group separately.

A Preview of the Chapters

The main objective of this book is to assess the health of science in America. However, an informative assessment is not a simple matter in this case, as "American science" is not a monolithic entity. Such an assessment requires multiple indicators that may reveal different trends. Understanding the nature of this heterogeneity is important, as this can help policymakers to design programs to promote American science more effectively. Therefore, in each chapter we examine American science from a different perspective, providing a more comprehensive view than can be achieved through inspection of only a single indicator.

In Chapter 1, we provide an overview of the development of science as an occupation and the evolution of American science since 1900. In Chapter 2, we shift from a historical overview to a spatial one, placing American science within the context of a globalized world and comparing it with science in other countries in terms of several different indicators. Chapter 3, a theoretical discussion of scientific career choice, provides the foundation for the rest of the book, which is divided into empirical studies of three types of trends in the United States since 1960: characteristics of the scientific labor force, societal support for science, and the supply of potential scientists. In Chapter 4, we study the demographic and economic characteristics of the American scientific labor force, presenting a portrait of its changing composition based on data from the U.S. Census over the period 1960–2000 and the American Community Survey in 2006–2008. In Chapter 5, we shift our focus to public attitudes, documenting trends in Americans' perceptions of and attitudes toward science since the 1970s. In Chapter 6, we describe trends in youths' expectations of scientific study. In Chapter 7, we assess trends in scientific education, examining scientific degree production and attainment at American institutions. In Chapter 8, we document trends in the transition from scientific education to scientific occupations. Finally, in the conclusion, we summarize our findings and explore their overall significance.

1

The Evolution of American Science

Science is a wonderful thing if one does not have
to earn one's living at it.

—ALBERT EINSTEIN, 1951

American science is a big industry that employs a big labor force. Here are some things we know about it:

(1) Science is huge. Estimates of the American science workforce vary based on how "scientist" is defined, but even the lowest estimates of scientific workers are in the millions.[1]

(2) Science involves enormous amounts of money. In 2006, the United States spent over $340 billion on research and development.[2] This includes significant spending on basic research, as evidenced by the National Science Foundation's budget of $6.87 billion in 2010 and the National Institutes of Health's budget of $31 billion the same year.[3]

(3) Science in the United States is conducted almost entirely by paid professionals. These scientists are employed by universities, private industry, nonprofit organizations, and government agencies and are generally well paid. In May 2006, according to the Occupational Employment Statistics Survey, the median annual income for those in science or engineering occupations was $64,160, more than twice the median ($30,400) for all occupations.[4]

(4) Science requires extensive training and high levels of skill. To be a professional scientist, one must typically have at least a bachelor's degree in one of the sciences. Among members of the scientific elite, a PhD is almost universal, most often from one of a small number of prestigious universities.[5]

These characteristics, central to science today, would be totally inaccurate in describing science 150 years ago or earlier. What was science like before the modern era? How did science evolve from what it used to be into what it is today? What are the consequences of this evolution for our study? In this chapter, we address these questions by providing a brief history of modern science as a profession. By examining the development of modern science, we document the incremental steps that led to American science today, with its features highlighted above. A theme that emerges from this review of history is that scientists have evolved gradually over time from amateurs into professionals. Only by understanding the evolution of science from its upper-class, amateur roots in a few European countries to the massive, worldwide, professional enterprise that exists today can we hope to make sense out of the values, ethics, and incentives that inform the career decisions of present-day scientists and prospective scientists, decisions that will collectively determine whether American science will continue to grow or fall into decline.

The European Roots of Professional Science

Relative to most other major institutions, such as government and the family, which have been part of human life almost from its beginning, modern science has a short history. It originated as part of "natural philosophy." As we will show in this section, early science could be considered a "profession" only insofar as it involved a community of practitioners with similar interests and shared codes of conduct.[6] Science could not be considered an occupation until the twentieth century, when, for the first time, a large community of individuals began to earn generous livelihoods as professional scientists. Before the twentieth century, however, the pursuit of science was generally beyond the means of all but wealthy gentlemen.[7]

By conventional accounts, modern science began with Nicolas Copernicus (1473–1543).[8] Earning his livelihood as a clerical administrator in his native Poland, Copernicus initiated a scientific "paradigm shift"[9]—a fundamental change in the assumptions governing the practice of science—by advancing the radical proposition that the sun, not the earth, was the center of the universe. This shift had enormous implications not only for astronomy and physics but for human knowledge in general, as it dispensed with the assumption that man occupied a privileged place in the universe. Following the publication of Copernicus's *De revolutionibus orbium coelestium* (On the Revolutions of the Heavenly Spheres) toward the end of his life, other astronomers such as Kepler and Galileo took up where he had left off, confirming and refining his theory with their observations, in a cumulative process called the Copernican Revolution, which was not completed until Newton published his work in the late 1600s.

Meanwhile, in Renaissance Italy, where Copernicus had gone to study, scientists were beginning to congregate and to form some of the earliest scientific societies. Joseph Ben-David argues that the evolution of scientific organizations played a significant role in promoting innovation and progress in science.[10] One of the earliest such influential organizations was the Academia dei Lincei (founded in 1603), to which Galileo belonged. Yet the membership of these organizations, as well as the pool of scientists in general, drew primarily from upper-class men with abstract thinking styles and aristocratic attitudes. Such scientists justified their work through appeals to Platonic philosophy rather than usefulness—a limitation that may have been responsible for the decline of Italian science by the seventeenth century, a period in which the center of scientific advancement shifted to England.[11]

One critical advancement that took place in England was the increasing differentiation of science from philosophy and all other forms of knowledge.[12] According to an idealized view espoused by the influential English philosopher Francis Bacon (1561–1626), science was the systematic investigation of law-like regularities in the natural world. During the seventeenth century, Bacon's "scientific method"—a phrase that took on multiple meanings as it gained acceptance—was used to mean objectivity, empiricism, and sometimes experiments.[13] With the rise of science as a social institution and the emergence of the scientific method as the preferred approach to understanding nature, "scientists" came into being as a select group of privileged men who practiced science as a superior mode of inquiry into nature in the English-speaking world.[14] The glory of English science was epitomized in Isaac Newton.[15] By showing that the motions of terrestrial and celestial objects were governed by the same set of natural laws, Newton's work in effect settled all theoretical questions raised by Copernican theory.[16]

While Newton earned his living as a professor of mathematics at Cambridge University, the vast majority of scientists living before the nineteenth century were amateurs and not what we would consider "professionals," as they did not earn their living by engaging in scientific pursuits. Early scientists were virtually all aristocrats, priests, doctors, or merchants and thus had independent sources of livelihood. They pursued science at their own expense, in search of truth and recognition. An example of a merchant who was also an amateur scientist was an American, an unlikely member of this nobleman-scientist group, Benjamin Franklin (1706–1790). A voracious reader, Franklin was an autodidact with little formal education. After having acquired wealth as a publisher, he chose to devote his life to politics and science.[17] Today Franklin is remembered not only as a framer of the U.S. Constitution but also as a scientist who discovered how electricity worked and who invented the Franklin stove, the lightning rod, bifocal glasses, and the urinary catheter.[18] While Franklin was able to pay his own scientific expenses, as were those with inherited wealth, scientists with no independent means sought sponsorship from the aristocracy or the Church. For example, Galileo relied on the patronage of princes and popes and received a high salary at the pinnacle of his career.[19]

The wealthy Darwin-Wedgwood family of England provides an interesting example of self-sponsored scientists. The family contained at least ten fellows of the Royal Society as well as several artists and poets.[20] The most prominent member of the family was Charles Darwin (1809–1882), who is known today for his theory of natural selection. Another well-known member of the Darwin-Wedgwood family was Charles Darwin's cousin Francis Galton (1822–1911), who laid the foundation of modern statistics and the use of statistical methods for the study of human phenomena in a variety of fields, such as psychology, sociology, demography, and anthropology.[21] Private wealth enabled family members such as Charles Darwin and Francis Galton to pursue science as an unpaid career.

The formation of scientific societies and academies was another important step toward the professionalization of scientists.[22] Typically, these academies did not provide any income for their members but provided places for the gathering of amateurs par excellence. The most notable example was the Royal Society, which proved to be instrumental for the rapid development of science in England.[23]

However, by the eighteenth century, England had lost its lead in world science, as France emerged as the new center.[24] The French counterpart to the Royal Society, the Académie des sciences, was established in 1666 under the auspices of the French state as an elite organization. Although members were paid for their labors, academicians were unlikely to pursue science as a full-time occupation, knowing that appointment to high-paying positions required a long waiting period.[25] Nevertheless, the practice of science in France was centralized in the scientific society and academy. The professionalization of science thus took root early in France, as sponsored scientists began to receive pay for their specialized services to the government.

Science in France began to stagnate and decline after 1830, as the centralization of the French system resulted in rigidity and the separation between education and research, which may have been barriers to further advancement. By the last decades of this century, the center of world science had shifted to Germany.[26] In contrast to England and France, where societies and academies were the dominant scientific institutions, universities were primary centers of scientific productivity in the German-speaking world.[27] In the German system, research became not only legitimate but indeed a vital activity of a university professor. The integration of science into universities in Germany was facilitated by a philosophical calling, distinctly German, for the intellectual unity between the scientific approach and the humanistic (or philosophical) approach to knowledge.[28] This development had two important ramifications: First, the number of potential university faculty members and students could be large, making it possible for science to become a paying profession. Second, science housed in universities could be flexible, competitive, cross-disciplinary, and free from external control. In sum, the most likely practitioners of science in Germany were professors, who began to be recognized as professionals.

Although the variation in space and time makes neat generalizations difficult, science practice and practitioners between the seventeenth and nineteenth centuries can nevertheless be characterized as largely amateurish in that few practicing scientists pursued science primarily as an occupation. Empirical studies of early scientists virtually all show that few of them practiced science for a living.[29] These early examples of "gentlemen scientists" stand in contrast to modern scientists, who can expect to support themselves financially through their scientific endeavors, in a development that has opened the practice of science to those from less advantaged backgrounds. Nonetheless, these early patterns of scientific practice have had enduring effects. Membership in professional organizations continues to be a marker of participation in the scientific community, and universities remain hubs of scientific research. While scientists may no longer justify their work on philosophical grounds, "science for science's sake" remains a tenet of basic scientific research, and concerns regarding the dependence of scientists on external entities for funding remain.

The Professionalization of Science in the United States

The professionalization of science was finally achieved in the United States in the twentieth century by meeting three requirements that we consider necessary for an activity to be considered a profession: (1) self-regulation or autonomy, (2) systematic training for qualification, and (3) payment for services.[30] As we have seen, pre-1900 science in Europe had not met all of these criteria. Although in Europe the evaluation of scientists' work or results was autonomous from influences external to science, there was little formal evaluation of scientists for paid positions, as key personnel decisions regarding scientists were made by persons external to science, such as political or religious authorities or wealthy patrons. Moreover, there was no explicit certification for specialized training in Europe comparable to that provided by the American graduate school. Finally, pre-1900 European scientists were not all paid to provide research as a specialized service.[31]

All this was to change in twentieth-century America, where science was finally transformed into a true profession.[32] To understand the historical context of this change, it is important to understand the developments under way both in the United States as a whole and within American science. The most important societal development was the rapid economic growth resulting from industrialization between 1860 and 1930. As measured by the gross domestic product (GDP), a common measure of economic output, the U.S. economy grew steadily at an annual rate of 4 percent between 1790 and 1930 after adjusting for inflation.[33]

Internally, the development of American science involved two separate but continually interwoven strands: (1) "pure science," in which the scientist's primary purpose is to understand the universe, and (2) "applied science," in which the goal is to create better products, methods, and technologies for the benefit of industry, consumers, government, the military, and other social groups. These two strands are associated with different scientific institutions.

"Pure science" developed primarily within American colleges and universities, in which research and training were conducted first in physical and only later in social and applied sciences. Two great universities were founded in America in the late nineteenth century: the Johns Hopkins University in 1876 and the University of Chicago in 1891. Both schools explicitly incorporated the German model, but with one important difference: they contained research-oriented graduate schools. Such schools served as models for many other American universities, including both private universities, such as Harvard and Yale, and, somewhat later, large public universities, such as the University of Michigan and the University of California.[34] By the early 1900s, the "professionally qualified research worker" was an established phenomenon, and a PhD in science was regarded as the equivalent to an MD.[35]

The strand of applied science owed much to America's long-standing love affair with technological innovation.[36] Colonists and settlers struggling to survive in harsh new wilderness environments often had to find creative ways of making do with whatever materials were at hand: hence the notorious "Yankee Ingenuity" Mark Twain celebrated in his novel *A Connecticut Yankee in King Arthur's Court*.[37] Rather than working to develop models of how the universe worked, American inventors such as Franklin, Bell, and Edison worked independently to solve practical problems, including fireproofing buildings from lightning, communicating over long distances, and illuminating houses after dark.[38] As early as 1641, colonial inventors began petitioning governments for patents, and when the American Constitution was framed in 1789, the federal government was empowered to create a patent system. Such a system allowed inventors to make handsome profits from their creative work and contributed to the unprecedented level of innovation in the late eighteenth and nineteenth centuries. New inventions included the automated flour mill, cotton gin, steamboat, telegraph, electric light bulb, and countless other world-changing devices.[39]

By the early twentieth century, the plethora of American inventions had led to the growth of factories and the gradual replacement of individual inventors with corporate research laboratories. Large manufacturing corporations began to employ scientists. By 1931, 1,600 U.S. companies had research laboratories.[40] Industrialization and mechanization changed social relations; labor itself was subject to scientific study for efficiency, as American ingenuity was applied to finding ways of organizing people to make larger numbers of things at an ever faster pace.[41] This can best be seen in Henry Ford's assembly line and Frederick Taylor's "scientific management," which ironically resulted in depriving workers of the freedom to exercise their own creative inventiveness.[42] The rapid growth of factories and the increasing reliance on machines during industrialization in America stimulated the demand for a wide variety of practical skills, which was met by the establishment in this period of professional schools and technical colleges, many of them land-grant colleges where students learned skills in teaching, agriculture, home economics, or medicine. Although such institutions were meant to produce professionals, training for applied research became part of the activities.

Despite the growth of engineering schools and corporate research labs, not all American inventors were willing to relinquish their autonomy to large institutions. Some of the inventors credited with the most radical breakthroughs chose to remain outside both the corporate and the academic worlds.[43] Such breakthroughs included the airplane by the Wright brothers in 1903, the Polaroid camera by Edwin Herbert Land in 1932, the photocopier by Chester F. Carlson in 1937, power steering by Francis W. Davis in 1951, and the computer mouse by Douglas Engelbart in 1968.[44]

As science and invention continued to evolve in America over the twentieth century, both industrial and academic scientists began to form connections with government agencies, particularly the military during the two World Wars and the Cold War. Government agencies provided funding to scientists for work that might potentially further the agencies' goals. The case of physicists who joined the war effort during World War I provides an early example of this phenomenon.[45] With the rise of Hitler in Germany, American science benefited from the influx of German-born and other European scientists who sought sanctuary from the Nazis. Albert Einstein (1879–1955), who came to America in 1933 to take a position at Princeton and never left, was the most celebrated member of this group.[46] The most obvious example of American scientists collaborating with the military was the Manhattan Project during World War II, during which a group of American scientists directed by physicist J. Robert Oppenheimer worked with the U.S. Army Corps of Engineers to develop the atomic bombs that were dropped on Hiroshima and Nagasaki, leading to Japan's surrender.[47]

In the aftermath of Hiroshima and Nagasaki, science faced a brief falling-out with the American public. The destructive capacity of the atomic bomb inspired fear and moral outrage, sentiments that were sometimes directed toward scientists, who were now viewed as enablers of destruction.[48] Despite their newfound ambivalence toward scientists, many Americans still interpreted the development of the atomic bomb as firm evidence of the nation's scientific and technological superiority. This optimistic attitude, however, was to be seriously challenged.

Science Race: A Tale of Two Nations

On October 4, 1957, the Soviet Union launched its first artificial satellite, Sputnik 1. The object itself was negligible: it weighed only 183 pounds, was described as "about the size of a grapefruit," contained only primitive instrumentation, and emitted nothing but beeps. Sputnik 1 was followed the next month by Sputnik 2—a larger satellite that contained a live dog. The tremendous significance of these two Soviet satellites in the history of American professional science cannot be overestimated, mainly because of their psychological impact. When, in his State of the Union address in January 2011, President Barack Obama used the words "Sputnik moment" to designate a historical point of extreme challenge, he was referring to the satellites that had a dramatic effect on Americans' attitudes toward science for half a century.[49]

The immediate reaction to the Sputnik launchings on the part of the American public in 1957 was a mixture of shock and excitement. Prior to Sputnik, the Soviet Union had been seen as a scientifically backward country whose military technology posed little threat to the United States, but now it could send an object flying over America. Although intercontinental ballistic missile (ICBM) technology was used only to launch a research satellite, Americans were all too aware that the Soviets might just as easily attach a nuclear warhead to a missile, a capability that the United States lacked at that time. "There can be no more underestimating Russia's scientific potential, either for war or peace," a *Chicago Daily News* editorial declared.[50] Senator Richard B. Russell, chairman of the Senate Armed Services Committee, asserted that the satellite launch confirmed that the Soviets had perfected an ICBM, which represented a serious new danger to the United States.[51] Thus, as a matter of national security, it seemed essential that American scientists and engineers act quickly to match the Soviets' successes to maintain the balance of power that had emerged between the two dominant nations after World War II.

While the Sputnik event intensified Americans' anxieties over the possibility of falling behind the Soviets in military power, it also opened up exciting new possibilities. If the Soviets could put a satellite into space, so could other countries. The American public, conditioned by the past experience of defeating Japan with a scientifically developed surprise of their own, expressed optimism that the United States would rise to this new challenge and would overtake the Soviets' early lead in the "space race." The new objective? Be the first to go to the moon. In a Gallup Poll conducted in Washington, DC, and Chicago a week to ten days after the Sputnik 1 launch, 61 percent expressed confidence that the next great achievement in space would be made by the United States, not the Soviet Union, with only 16 percent betting on the Soviet Union.[52] Public excitement about mankind's entry into outer space was accompanied, according to the *New York Times*, by increased sales of telescopes and binoculars, while a survey of U.S. managing editors conducted by New York University in 1958 reported a 50 percent or greater increase in the amount of space given science in their newspapers since the Sputnik event.[53]

Meanwhile, politicians were quick to capitalize on the public reaction to Sputnik, dividing along party lines between Democrats, who blamed the Eisenhower administration for policies that slowed scientific efforts and for underfunding scientific research, and Republicans, who minimized the significance of the launch while simultaneously taking corrective actions. Although President Eisenhower declared on October 9 that the Sputnik launch had not raised his anxiety regarding U.S. security "one iota," a few weeks later he created the position of special assistant for science and technology and named MIT President James R. Killian to the post. Killian created a "President's Scientific Advisory Committee," which developed plans for the National Aeronautics and Space Administration. In less than a year, Congress had passed the National Aeronautics and Space Act creating the agency, and on July 29, 1958, the president signed it into law.

Bolstering the American space program was only part of the U.S. government's response to Sputnik. It also wanted to make sure that no new technological breakthrough in another country would ever again take the United States by surprise. The result was a second agency to fund relevant research, the Advanced Research Projects Agency (ARPA), the purpose of which was to "prevent strategic surprise from negatively impacting U.S. national security and create strategic surprise for U.S. adversaries by maintaining the technological superiority of the U.S. military."[54] So strong was the political commitment to science following Sputnik that the president and Congress, as well as their successors for several decades to come, were able to think far beyond the needs of the immediate present. For a few decades, no price seemed too high for victory in the new race for scientific superiority. In addition to recommending a massive push in research funding, President Eisenhower's science advisors urged him to emphasize science education and scientific careers. Speaking to this effect, Eisenhower asked the American public to put science education "above all other immediate tasks of producing missiles, of developing new techniques in the Armed Services," thus framing the crisis as a challenge in education rather than simply a challenge of military technology.[55] Such statements were followed by large increases in public school funding aimed at providing a solid educational foundation for the next generation of physicists and engineers.

The sentiment that the United States was in a major science race with the Soviet Union was frequently echoed at the time. For example, in his newspaper column in the *New York Herald Tribune* on October 10, 1957, famed journalist Walter Lippman wrote,

> This is a grim business. It is grim, in my mind at least, not because I think the Soviets have such a lead in the race of armaments that we may be soon at their mercy. Not at all. It is a grim business because a society cannot stand still. If it loses the momentum of its own progress, it will deteriorate and decline, lacking purpose and losing confidence in itself.[56]

A few months later, in January 1958, American physicist and engineer Lloyd Berkner wrote,

> The achievement of the Soviet satellite has demonstrated to Americans what they refused to believe before, that they are in a race for intellectual leadership when they hadn't realized that there was a race, or even that another nation had the capability to challenge their technology.[57]

The underlying mindset of such remarks was clear: science is a race between nations that any given nation loses at its own peril. The mission of the scientist is not the expansion of human knowledge for its own sake but the development and safeguarding of intellectual weapons needed for national security and economic strength. Without science, society is doomed to decline. This represents a significant departure from the conception of science prior to the twentieth century—as an enterprise undertaken largely by self-funded individuals in pursuit of knowledge (and sometimes profit). The Sputnik event framed science and scientific progress as a critical factor in a nation's security and political power and encouraged Americans to view science as part of the proving ground for competition among nations.

As the legacy of Sputnik, this mindset has continued to the present day, informing as well as framing the current discourse on America's potential failure to "compete" successfully with other nations in scientific arenas. In assessing the validity of the alarmists' claims, we should remind ourselves of the history that gave rise to this discourse. Interestingly, while much of the discussion on American science is framed in terms of international competition rather than international collaboration, the U.S.-Soviet space race within which the competition paradigm originated has long since outgrown it, as Americans and Russians now engage in joint missions, and astronauts and scientists from multiple nations inhabit an International Space Station. If the space race is now irrelevant after the end of the Cold War, one wonders whether the science race as a metaphor is still applicable today.

From Little Science to Big Science

We have seen how, through the two crucial historical events of first Hiroshima/ Nagasaki and then Sputnik, science became fully, tightly, and irreversibly integrated with national interests, both military and economic, in the United States. As a result, science funding increased dramatically. So did the size of science as a form of human activity. Gone forever was the Little Science practiced by a handful of amateurs in the past; in its place, Big Science, conducted by armies of professionals, had been born and was here to stay.

Public support for U.S. science increased rapidly during World War II and the Cold War, as did federal funding. The federal government's share of total research and development expenditures increased rapidly from less than one-quarter in 1940 to more than two-thirds in 1965.[58] Two agencies became massive distributors of federal funds during this period. Established in 1950 to counter the increasing role of the military in academic science during World War II, the National Science Foundation (NSF) provides federal support for education and fundamental research in the sciences and engineering.[59] Its budget for research support grew from $3.5 million in 1952 to $6.87 billion in 2010.[60] Another important government agency supporting science in the United States is the National Institutes of Health (NIH), which was created from other federal agencies in 1930. NIH began a program of extramural research grants and fellowship awards in 1946. Its congressional appropriation increased rapidly from about $500,000 in 1946 to over $31 billion in 2010.[61] This large investment in science, including that for science education, yielded a high rate of growth in the scientific output of the United States during the post–World War II years.

Even as Americans were landing on the moon in 1969, however, science was becoming the focus of increasing criticism and disenchantment. In his 1971 book *The Physicists: The History of a Scientific Community in Modern America*, Daniel J. Kevles attacked the scientific profession as elitist and hierarchically organized, though he also recognized that elitism may be consistent with the "ethic of disinterested science."[62] Critics of science blamed it for the nuclear arms race and abuses of military power during the Vietnam War, both of which were seen as the result of new research products.[63] For the American public, twenty-five years of involvement of science with military endeavors was becoming increasingly problematic.

By the 1970s, postmodern philosophers and some scientists began to seriously question the notion of scientific objectivity. Attention in science studies now shifted away from scientific activities to the effects of gender, age, ethnicity, religion, class, or other defining characteristics on scientific projects and outcomes.[64] In recent decades, the underrepresentation of women in the sciences has been a major topic, because such underrepresentation has persisted even as women have made significant advances in educational attainment and entrance into other male-dominated professions.[65] Besides concerns about social justice and the potential waste of talent, many scholars now consider how the dominance of white and Asian men in science may have shaped the formation of scientific inquiries and the interpretations of study results, possibly leading to disenchantment and low levels of participation in science among women and other racial and ethnic minorities.[66]

With the federal government's continued and intensive investment in science, the "Little Science" of the past has been transformed gradually into the "Big Science" of the present.[67] In contrast with Little Science, Big Science is characterized by very large government expenditures, which support an enormous scientific labor force and have resulted in rapid growth in scientific productivity. In consequence, growth in scientific funding far outpaced overall economic growth in the last century. For example, the per-capita cost of supporting science increased at the annual rate of 15 percent between 1950 and 1960 in the United States, while the GDP expanded at only 3.5 percent per year.[68] Recognizing the power of this rapid exponential growth, Derek Price predicted its inevitable limits. "If we did [continue the growth of science as in the past]," he wrote, "we should have two scientists for every man, woman, child, and dog in the population, and we should spend on them twice as much money as we had."[69]

Was the rapid development of science during the period immediately following World War II an atypical phase of U.S. history that can no longer be sustained today? Have interest in and public support for science waned in recent decades? Has the experience of the Cold War soured Americans on the role of science in military force? Has the focus on Big Science in contemporary America led to overcentralization and stagnation? In the rest of the book, we will attempt to answer these questions.

Scientists: Humans or Numbers?

One important consequence of the transition from Little Science to Big Science involves the perception of scientists. No longer seen as persons of "peculiar characteristics" pursuing an unusual hobby, scientists are now regarded as "relatively normal people," who are "just perhaps more intelligent."[70] While scientists may still be motivated by a quest for knowledge, Big Science makes it possible for many individuals to be attracted to science for rewards in terms of money and social status. In other words, today's scientists are trained, regulated, and paid as professionals.[71] Though still respected for their intelligence and industry, scientists may now be thought of as ordinary professionals rather than as upper-class amateurs pursuing noble causes. Hence, society must provide both the resources and the motivations to scientists, who are no longer motivated exclusively by their innate curiosity and funded primarily by their own resources. In later chapters, we will explore the implications of this change for the recruitment of scientists, documenting why many potential scientists may have mundane reasons for leaving science at different stages of their careers.

As seen above, modern science originated as a pursuit of European noblemen and the independently wealthy. Early scientists were either gentlemen themselves or sponsored by the upper classes.[72] To be sure, not every gentleman who dabbled in physics or biology proved to be a Newton or a Darwin, just as not every contemporary PhD will win the Nobel Prize. For many early practitioners, science was probably nothing more than an enjoyable hobby, on a par with fox hunting or art collecting. For others, however, science became a serious vocation, and a handful of these individuals laid the groundwork for the various branches of scientific inquiry. At a time when the theoretical foundations of science were in their infancy, it was perhaps easier for exceptional individuals to distinguish themselves from their peers than in the age of Big Science.[73]

The aristocratic origins of science may partially explain why, in an age in which science is just another profession, scientists are motivated not only by economic rewards but also by the desire to be recognized as exceptional and by their devotion to finding truth in nature. Although they occupy an elite position within society, scientists, particularly academic scientists, now compete not only for higher pay but also for elite status. For the most ambitious scientists, a key motivation may be the desire to win recognition, in such forms as prestigious awards, publication in leading journals, appointments at highly respected institutions, memberships in science academies, and the attention of the research community. As is well known, the inequality of rewards within science is extremely large, with only a minority able to become members of the scientific elite, who dominate the top journals, hold choice positions, and win top awards.[74] At the top of the scientific hierarchy, there are "great scientists" who are recognized by all in the same field as individual humans. The rest are ordinary scientists whose existence and contributions are, to put it bluntly, more or less counted as parts of numbers.

There are many signs that "great scientists" are still with us today, as they were in the past. Price pointed to the emergence of groups of elite scientists as separate entities from the mass of less distinguished peers. These groups generally contain no more than one hundred members in a particular discipline.

Their members are continually communicating with one another (by the circulation of preprints in the past, now over the Internet) and visiting each other so frequently as to create "invisible colleges."[75]

Exceptionally creative individuals account for groundbreaking discoveries, for which they may receive a Nobel Prize or be elected to a science academy. However, to a larger extent than before, such individuals do so not as solitary pioneers but rather as leaders of large scientific teams; they can thus be seen as the descendents of Little Science armed with the tools and supports of Big Science. In contrast to their predecessors who stood alone, the outstanding scientists of the present are supported by teams of scientific workers whose contributions may be virtually interchangeable but nevertheless necessary. Such a situation creates a long-standing tension between scientists as unique individuals and scientists as numbers, a tension that has existed since science first became professionalized. Today, all scientists may need to decide how much effort to devote to trying to distinguish themselves as opposed to being a supportive community member. Likewise, science administrators may feel torn between hiring stars versus hiring "good team players."

The professionalization of science in the twentieth century not only replaced Little Science with Big Science but also transformed scientists from amateurs to professionals who could be recognized in their social roles both as individual humans and as numbers. As a result, science can be thought of as a job like any other that requires postsecondary education and, in return, rewards workers with above-average salaries and social prestige. This transformation may encourage some individuals to pursue science who could not have done so when the discipline was largely an unpaid activity. At the same time, Big Science may make a scientific career less appealing to those individuals who might be attracted to the life of solitary work but considerable prestige that Little Science used to offer in earlier periods. In other words, Big Science has made science simultaneously more secure for those who choose science because of its work characteristics and more of a gamble for those who wish to achieve fame and power through individual achievement.

Summary

In our brief review of the history of science in this chapter, we established that the world center of science has shifted from one country to another in the past. We showed how modern science was first developed in Europe, its center shifting from Renaissance Italy to England in the seventeenth century, then to France in the eighteenth century, and lastly to Germany in the nineteenth century before crossing the Atlantic to the United States. Each time the world center of science shifted, important new elements were introduced:

(1) The appreciation of scientific work among elite academic societies began in Renaissance Italy.
(2) The evaluation of scientific work as "truths," independent of individuals' subjectivities, was a key element contributing to the separation of science as a privileged mode of inquiry in seventeenth-century England.
(3) The success of science in eighteenth-century France owed something to the direct sponsorship of the state.
(4) The Germans surpassed the French in scientific leadership by relocating the institutional basis of science within universities and by integrating knowledge across disciplines in the nineteenth century.
(5) Science did not fully achieve professionalization until the twentieth century in the United States. Through industrialization, World War II, the Cold War, and the space race, science in the United States came to be viewed as integral to the nation's economic strength, military power, and public well-being in general, and consequently, it grew much larger than ever before, in terms of both funding and personnel. Huge sums of money have been poured into American science, and huge numbers of people have worked as scientists.

Although the world center of science has shifted from country to country in the past, it would be premature to infer from this historical pattern that the world center of science will soon shift away from the United States in the near future. A major reason is that today's science is qualitatively different, having been transformed from Little into Big, that is, from individuals' unpaid pursuits into government-sponsored paid services. In the course of this transformation, scientists changed from amateurs into professionals who were increasingly regarded as numbers rather than as human beings. This means that today's science is affected much more by government policies than was the case in the past. It also means that, as paid professionals existing in large numbers, today's scientists are affected more by incentives such as financial rewards than was the case in the past.

The characterization of today's scientists as both human beings and numbers presents a dilemma for us in studying contemporary American scientists. We know that, at a fundamental level, scientific activities are accomplished by individual scientists who are all unique in their own contributions to science. However, we are social scientists relying on quantitative data as evidence for research and resort primarily to quantitative information in this book on American scientists. To some extent, this focus on quantitative information is justified in light of today's Big Science, in which more-or-less interchangeable scientists can be studied by numbers in the aggregate. However, we are keenly aware of a serious limitation of this approach, as we know that science has been, and will always be, practiced by individuals. Although the funding model or the institutional arrangement for scientists of the present represents a radical departure from those of the past, the ideal of Little Science, with elite scientists at its core, is very much alive today.

We believe Big Science is worth studying, but we also recognize that the picture we present in this book can be only partial. Any statistical analysis is based on an implicit assumption about homogeneity, which in our case means that one scientist (or one scientific product) in a particular category is more or less the equivalent of another scientist (or another scientific product) in the same category, as defined by some measurable characteristics. In reality, we know that this assumption is not only too simplistic but wrong, for it undermines a basic premise in science—that every scientist is unique. As unique individuals, scientists are not all interchangeable and thus not subject to statistical analyses. As members of a large profession, however, scientists, like members of other professions, exhibit certain defining characteristics that can profitably be studied as numbers. While we make extensive use of quantitative information about scientists in this book, we acknowledge the methodological limitation of our approach.

2

American Science and Globalization

> Patterns in the rise and fall of former leading scientific nations imply
> that, unless serious steps are taken, the United States could look back on
> the early twenty-first century as the peak of its scientific dominance. . . .
> Each former scientific power, especially during the initial stages of
> decline, had the illusion that its system was performing better than it
> was, overestimating its strength and underestimating innovation
> elsewhere. The elite could not imagine that the centre would shift.
>
> —J. ROGERS HOLLINGSWORTH, KARL H. MÜLLER, AND
> ELLEN JANE HOLLINGSWORTH, 2008

As shown in the last chapter, the world center of modern science has shifted from one country to another over modern history, from Italy to England to France to Germany and, finally, to the United States. In the past, each of these centers of science has enjoyed world leadership for only about eighty years.[1] The American dominance of science, however, has now lasted for more than ninety years, from the 1920s to the present, prompting some observers to raise the question of whether this dominance is in serious jeopardy and whether America may be guilty of the same overconfidence that brought about the scientific downfalls of other nations in the past.[2]

Those commentators who are worried about American science do not, of course, base their reasoning exclusively on cyclic historical patterns in world centers of science. As we argued in the last chapter, the historical pattern may no longer repeat itself, as no science-dominant European country was ever comparable to the contemporary United States in the sheer scale of its scientific institutions and achievements. Big Science is a uniquely American product that is unprecedented anywhere else throughout world history.

In addition, no previous center of world science has been operating in a globalized world in which national boundaries, across which information is now freely exchanged, have come to mean less and less. Although we may still debate its concrete manifestations, specific effects, and desirability, there is little doubt that globalization—interconnectedness across the globe—has been taking place in the most fundamental ways in recent years.[3] Speculations about the reasons for globalization abound, including the breakdown of the former Soviet bloc, inexpensive telecommunication and transportation, dominance of U.S. culture and the English language, and, of course, the development of computer and Internet technology.[4] In a globalized world, one may well ask whether the old rules still apply. As such a world has become increasingly interdependent, how has this changed the meaning of competition between nations for scientific expertise? Is scientific dominance still going to be possible for any one country, or is it destined to be replaced by a leaderless global network?

As globalization progresses, it makes less and less sense to examine the progress of science inside the United States—as we shall do in Chapters 4–8—without first locating it within its global contexts, asking what external forces may be impacting the progress of American science. Overall, the impact of globalization for American scientists appears to be twofold: there have been both increases in collaboration and increases in competition. In this chapter, we will take a look at both of these before turning our attention, in the remainder of the book, to the growth or decline of science within America's borders.

Globalization and Collaboration

Unlike many other aspects of world culture, science—at least basic science—has been a global enterprise from its very inception, due to several unique features. First, at least for academic researchers, scientific products have long appeared in the public domain and been made available in international journals, to be shared, assimilated, and improved upon by anyone in any country. Consumption of scientific products by one researcher in one country does not reduce the likelihood of its consumption by another researcher in another country.[5] Second, evaluation of scientific contributions and achievements has always been based on more-or-less universalistic criteria, assumed to be unrelated to the scientist's personal attributes, including race and nationality.[6] Third, to a much larger extent than in other fields, the validity of scientific knowledge is universal, transcending boundaries of language, culture, political system, and religion. Of course, this universality is far from perfect.[7] Nevertheless, scientific knowledge is still universal enough to explain why it is relatively easy for foreign students who have received quality undergraduate science education in their home countries to compete effectively in U.S. science graduate programs and later in the U.S. scientific labor market.[8]

Science today increasingly requires collaborative efforts across national boundaries. A much higher proportion of U.S. scientists now collaborate with scientists in other countries than previously.[9] Worldwide, between 1988 and 2005, the share of publications with authors from multiple countries increased from 8 percent to 20 percent.[10] Such collaborations between individual scientists selected on the basis of their expertise rather than their geographic location facilitate scientific achievements that benefit the entire world. In the last ten thousand years, human societies have, for the most part, been able to make significant advances in economy and culture only when technological knowledge was shared over different regions.[11] Thus, rather than necessarily harming science in any one country, globalization may actually benefit scientific progress broadly.

One way to understand this is to consider whether science could be practiced exclusively within national boundaries. Even if this were possible, it would be hugely counterproductive, as science is cumulative in nature: scientists learn from knowledge produced by other scientists in order to contribute incrementally new knowledge. It is not possible for each group of humans to reinvent everything and make new progress. In this sense, Newton's famous line, "If I have seen a little further it is by standing on the shoulders of Giants," is really an early statement about the need for globalization in science.[12] Indeed, all the major scientists whose contributions were essential to the Scientific Revolution came from different countries: Copernicus was Polish, Kepler German, Galileo Italian, and Newton English.[13]

There is also ample evidence that innovative ideas are most likely to emerge when people with different perspectives interact with each other.[14] Scientists from different countries often have different experiences and perspectives, and this diversity can make international collaboration highly productive for all. Finally, as with trading, comparative advantages can be exchanged for mutual benefits between scientists in different countries.[15] Some economists even suggest that the United States may be better off, meaning more economically efficient, by "buying" scientists from abroad rather than training its own scientists.[16]

Globalization and Competition

While science is collaborative, it is also competitive on the micro as well as the macro levels. On the micro level, individuals compete for admission to academic science programs, for academic and nonacademic jobs as scientists, and, as working scientists, for publication, research grants, promotion, and awards. Globalization has the potential to affect the outcomes of this competition by increasing immigration opportunities for scientists from one country to another. On the macro level, political leaders and the general public tend to regard their nation's scientists as a precious resource that can make the difference between winning and losing the battle for economic and/or military supremacy. This perspective leads to public concern about whether American science is "in decline," based on cross-national comparisons in terms of students' math and science test scores, published papers, patent applications, Nobel prize winners, and other evidence of the quality of both prospective and working scientists. As examples of such cross-national comparisons, let us look at two indicators: students' math and science test scores and scientists' published papers.

MATH AND SCIENCE TEST SCORES.   One cause often cited for the relative stagnation of U.S. science is inadequate education in science among contemporary American youth. For example, the 2007 NAS report *Rising above the Gathering Storm* expressed concern about science education in the United States today lagging behind that in other countries. Here, globalization is at work indirectly. Globalization results in fast and virtually costless distribution of scientific advancements—including those in technology and medicine—across the globe, from more developed countries where such advancements were originally made to less developed countries where the advancements are now consumed and disseminated, facilitating economic development worldwide. Economic development in less developed countries is in turn associated with improvements in their education systems, which lead to a more productive and competitive labor force and thus to greater scientific achievements in these countries.

The NAS report maintained that U.S. students performed poorly relative to those in other countries on international tests such as Trends in International Mathematics and Science Study (TIMSS) and the Programme for International Student Assessment (PISA).[17] More recently, the release of the 2009 PISA scores in December 2010 brought a fresh round of discussion of the performance of U.S. students, as American teens performed slightly below the OECD average in math and at the average in science, while students in Shanghai, China, topped the charts in all three of PISA's content areas—reading, science, and math.[18] Secretary of Education Arne Duncan implored Americans to see these results as a "wake-up call" and raised concerns that Americans are being "out-educated" and at risk for becoming noncompetitive in a global economy.[19] For our own analysis, we use scores from the 2006 version of PISA and fourth graders' scores from the 2007 version of TIMSS; fifty-seven countries participated in the 2006 PISA and thirty-six countries in the fourth grade 2007 TIMSS.[20]

Both TIMSS and PISA are designed to assess the performance of school-aged children in multiple fields of study, including math and science, but the two tests have several key differences. The sample for PISA is age-based, while the sample for TIMSS is grade-based. Whereas TIMSS targets fourth and eighth graders, regardless of their age, PISA covers fifteen-year-olds, regardless of their grade. Variation in schooling practices by country will therefore lead to incomparability by country in the age of TIMSS-takers and in the grade levels of PISA-takers. Neither system is inherently better, as each test compares students from different countries that are similar in terms of either age or grade level, while allowing the other to vary. PISA and TIMSS also differ in the countries that participate. American students, therefore, may perform relatively better on one test than another in part because of a change in the comparison group. Last, PISA and TIMSS have content differences. TIMSS places greater emphasis on concrete skills and knowledge, whereas PISA aims to engage students' problem-solving and reasoning skills.[21]

In comparing countries in terms of educational outcomes, we need to control for a country's level of economic development, as more economic resources should lead to higher levels of educational achievement. For this reason, we first compare fourth graders' average TIMSS scores in 2007 against per-capita GDP in the same year through scatterplots and correlation analyses. To measure economic resources across countries in comparable terms, we use the Penn World Table Version 6.3, which adjusts for differences in purchasing power across countries.[22] The results for math are shown in Figure 2.1a, and those for science are shown in Figure 2.1b. Note that each dot in the figures represents a country/place, with those for the United States and several other notable countries delineated.[23] In each figure, we present the lowess (locally weighted scatterplot smoothing) regression line that fits the data best in describing the relationship between the TIMSS scores and GDP.[24] For both math and science, the line of best fit shows that increasing financial resources are associated initially with sharp increases in test scores, but this is followed by a substantial flattening out, with further increases in per-capita GDP leading to little additional gain in test scores.

The United States falls slightly below the line in math and slightly above the line in science, about as would be expected given the economic prosperity of the United States.[25] Hong Kong and Singapore, which have a GDP per capita similar to that of the United States, are above the line in both math and science, especially in math. In comparison to these high-performing countries/regions, American youth's performance appears poor.

Results from PISA, shown in Figure 2.2a and 2.2b, are somewhat less favorable.[26] Using the same procedure, we observe that U.S. students fall below the line of best fit in both math and science, although the disparity is greater in math. The top spot among this group of countries/places in 2006 went to Chinese Taipei in math and Finland in science. In conclusion, a general pattern that emerges from these figures is one of American students' mediocre performance, even though they have access to considerable economic resources in the United States.

The concern about American youth's poor performance in math and science relative to that in selected other countries, however, should be distinguished from the concern that American youth are not performing as well as in the

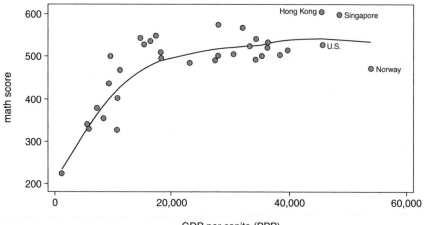

Figure 2.1a. TIMSS fourth-grade math scores by gross domestic product (GDP) per capita purchasing power parity (PPP), 2007.
Note: GDP per capita is adjusted for PPP using the *Penn World Table Version 6.3.* The GDP of the United Kingdom is used for England and Scotland. Kuwait is an outlier and removed from the analysis. Line of best fit is generated by lowess regression with bandwidth 0.8.
Sources: Gonzales et al. (2009); Heston et al. (2009); United Nations (2009).

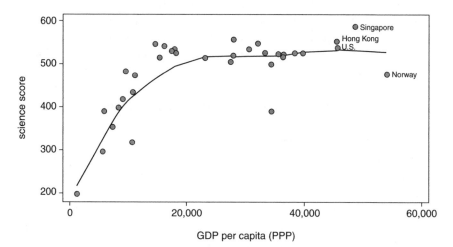

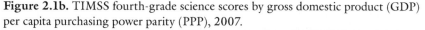

Figure 2.1b. TIMSS fourth-grade science scores by gross domestic product (GDP) per capita purchasing power parity (PPP), 2007.
Note: GDP per capita is adjusted for PPP using the *Penn World Table Version 6.3.* The GDP of the United Kingdom is used for England and Scotland. Kuwait is an outlier and removed from the analysis. Line of best fit is generated by lowess regression with bandwidth 0.8.
Sources: Gonzales et al. (2008); Heston et al. (2009); United Nations (2009).

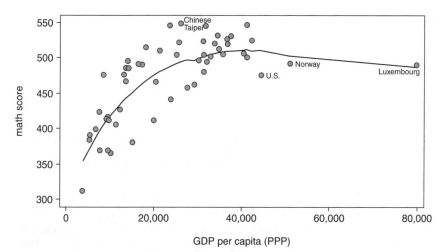

Figure 2.2a. PISA math scores by gross domestic product (GDP) per capita purchasing power parity (PPP), 2006.
Note: GDP per capita is adjusted for PPP using the *Penn World Table Version 6.3*. Serbia and Liechtenstein were excluded from the analysis because their GDP per capita for 2007 was not available from the *Penn World Table Version 6.3*. Kuwait and Qatar are outliers and removed from the analysis. Line of best fit is generated by lowess regression with bandwidth 0.8.
Sources: Gonzales et al. (2008); Heston et al. (2009); United Nations (2009).

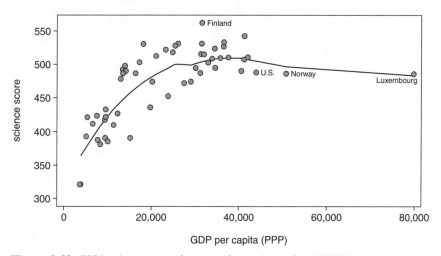

Figure 2.2b. PISA science scores by gross domestic product (GDP) per capita purchasing power parity (PPP), 2006.
Note: GDP per capita is adjusted for PPP using the *Penn World Table Version 6.3*. Serbia and Liechtenstein were excluded from the analysis because their GDP per capita for 2007 was not available from the *Penn World Table Version 6.3*. Kuwait and Qatar are outliers and removed from the analysis. Line of best fit is generated by lowess regression with bandwidth 0.8.
Sources: Gonzales et al. (2008); Heston et al. (2009); United Nations (2009).

past. Although the TIMSS began only in 1995, the scores of American fourth and eighth graders improved in math and remained steady in science between 1995 and 2007.[27] PISA has an even shorter time window, with math scores available only in 2003, 2006, and 2009 and science scores only in 2006 and 2009. However, changes in the measures for America were insignificant or positive over this period.[28] Therefore, there is no evidence that American students' performance has declined in these areas in recent years.

SCIENTISTS' PUBLICATION RATES. As the second indicator, we look at scientific publication rates to measure trends in scientific output. In column 1 of Table 2.1, we present annual growth rates in scientific article production in the United States, for four major science fields, over the period 1988–1992 and the period 1992–2003.[29] Two patterns emerge from these numbers. First, with the exception of physics in 1988–1992, growth rates in scientific productivity in the United States in the years after 1988 fall short of historical growth patterns in scientific output reported by science historian Derek Price in 1963 as being around 5–7 percent per year.[30] Second, the growth rates are lower in the more recent period, 1992–2003, than in the earlier period, 1988–1992, with the single exception of mathematics. For example, in physics, the growth rate was 5.1 percent per year in the United States in 1988–1992 but dropped to 0.3 percent per year in 1992–2003. A similar but less drastic trend occurred in chemistry in the United States (4.2 percent to 1.2 percent) and biology (1.7 percent to 1.1 percent).[31]

Table 2.1. Average annual growth rate in science article output, by country/ region, field, and year (percent)

	United States	EU-15	Japan	East Asia-4
Biology				
1988–1992	1.7	6.4	4.6	17.7
1992–2003	1.1	4.1	3.9	16.0
Chemistry				
1988–1992	4.2	5.7	6.6	33.3
1992–2003	1.2	2.3	2.4	16.1
Physics				
1988–1992	5.1	10.6	10.9	19.7
1992–2003	0.3	3.4	4.4	14.3
Mathematics				
1988–1992	–2.0	3.2	–8.1	18.1
1992–2003	1.4	6.7	8.0	14.2

Source: National Science Foundation (2007).

Taken together, we observe a slowdown in scientific output in recent years in the United States. Particularly noticeable is the fact that the highest rate of growth during the 1992–2003 period is only 1.4 percent. As discussed in Chapter 1, this slowdown was anticipated by Price, who believed that scientific growth should follow logistic rather than exponential growth patterns in the long run. A logistic growth pattern is one in which the growth rate is curtailed after a period of rapid growth due to the effects of saturation.[32]

If we are witnessing saturation, it is not yet the saturation of science per se but the saturation of American science. The recent U.S. slowdown in scientific output is particularly striking in comparison to the trend data in scientific output in other countries. We present the comparative data in the other three columns in Table 2.1, for EU-15 (member states of the European Union prior to May 1, 2004),[33] Japan, and East Asia-4 (China, Singapore, South Korea, and Taiwan).[34] In physics, chemistry, and biology, each country or group of countries experienced lower growth in the later period than in the earlier period, so the United States is not alone in experiencing a slowdown in growth. However, the U.S. growth rates are uniformly the lowest in both periods in all fields, with the single exception of mathematics in the 1988–1992 period. While the highest rate of growth in the 1992–2003 period in the United States is only 1.4 percent, East Asia-4 experienced double-digit growth rates in each field during the same period. Although the United States' high initial levels of productivity make such rapid growth more challenging, these results suggest that the gap between the United States and other countries in research productivity has been narrowing.

Thus, in recent decades, U.S. dominance in terms of research productivity has eroded to some extent. By almost all major indicators, the U.S. share of scientific output in the world has been declining since 1992 in almost all fields.[35] Regarding the top 1 percent cited articles in all science and engineering, for example, the U.S. share steadily declined from 64.6 percent in 1992 to 56.6 percent in 2003. This loss in U.S. advantage resulted in gains for European countries (from 23.3 percent to 27.7 percent), Japan (from 4.2 percent to 5.3 percent), other East Asian countries (from 0.1 percent to 1.1 percent), and all other countries (from 7.8 percent to 9.3 percent).[36] Within journals indexed by Thomson Reuters in the *Web of Science* (which contains articles in the social sciences as well as the hard sciences), the share of all articles with at least one U.S. coauthor fell from almost 40 percent in 1981 to 29 percent in 2009. As a result, by 2009, the European Union and Asia claimed higher shares of authorship for scientific articles than the United States (at 36 percent and 31 percent, respectively).[37] There is no question that, at least in terms of science publications, the rest of the world is catching up with America. Globalization has contributed to this trend.

Collectively, as we shall see, such cross-national comparisons as those conducted in this chapter, combined with other studies showing increasing proportions of immigrant scientists, indicate that America may, in fact, be in danger of losing its dominant position in the scientific world. This does not necessarily mean, however, that U.S. science is in decline. America's potential loss of dominance could simply be due to the fact that as a result of globalization, professional science in other countries has improved. America's situation would be much more precarious if science is indeed declining in America in an absolute sense than if science is merely getting better elsewhere. Thus, along with making cross-national comparisons, we also need to look at how current American science is doing relative to its own past performance. In other words, we need to take a historical approach to our subject, not just a cross-national one. This will be our task in the rest of this book.

Once we have completed our examination of how American science has changed in a number of different respects since the 1960s, we will revisit, in our conclusion, the topic of globalization. Having identified patterns of change within American science, our final task will be to ask what role globalization may have played in effecting these changes.

Summary

American science exists not in a vacuum but in a world that is becoming increasingly interconnected, that is, globalized. Thus, before embarking on a historical analysis of science within the United States, it is important to place this study within its international context and acknowledge those external forces that may be impacting the course of science within the country. In general, the forces of globalization seem likely to have two possible types of effects:

(1) Collaborative effects: Science has always been globalized, due to its widespread sharing of information, its universalistic criteria for evaluation, and the universality of scientific knowledge. Never before, however, have American scientists been able to communicate so easily with colleagues anywhere in the world. Consequently, the amount of international collaboration has dramatically increased, to the benefit of all.
(2) Competitive effects: On the micro level, individual scientists and prospective scientists compete for various types of rewards. Globalization also means increases in immigration of scientists to the United States, which may affect the distribution of such rewards. On the macro level, political leaders and others consider their nation's scientists a precious national resource for ensuring economic prosperity and

military power. Numerous cross-national studies have been done in response to concerns that America might be "losing the race" where science is concerned, with varying results. As examples, we looked at two arenas of global scientific competition: student test scores in math and science and the quantity of scientific papers published. Our results and those of other studies indicate that American science may be in danger of losing its dominant position in the world.

Loss of dominance, however, is not the same thing as decline. As White House Science Advisor John Holdren recently said: "We can't expect to be number one in everything indefinitely."[38] If America loses its status as the dominant leader of the scientific world, this could merely mean that science in other countries is getting better, not necessarily that American science is getting worse. Thus, cross-national studies cannot tell us whether American science is likely to advance or decline. A multifaceted, historical view of American science relative to its own past, however, can give us a sensible understanding of its progress or decline. The rest of the book is devoted to this task.

3

Why Do People Become Scientists?

One enters civil society by mere birth, and one becomes a citizen by mere coming of age. Not so in the Republic of Science, wherein membership must be diligently sought and selectively granted.

—BERTRAND DE JOUVENEL (QUOTED BY DANIEL BELL, 1976)

A shortage of scientists in a society is the collective result of not enough people making decisions to become—and remain—scientists. Thus, if we want to know whether a shortage is likely to occur, as the alarmists have claimed, it is important to understand how individual career decisions are made, particularly those regarding whether to pursue science. Toward this end, we will need to understand scientists not as an isolated group but as existing within a broader social context—taking into account such factors as educational institutions, wages of scientists relative to those of other professionals, public attitudes toward science, and social inequality in terms of social origin, race, and gender. In this chapter, we will look at some of the factors likely to influence individuals' decisions to pursue scientific careers.

While individuals make their own choices concerning whether to pursue a scientific career, such choices always have consequences for society at large. As stated in the Introduction, a career choice to become a scientist generates positive externalities: there are benefits to society when an individual chooses a scientific career. Conversely, career choices at the level of the individual are influenced by the larger social context. Thus, we will need to form bridges between the micro and macro levels in considering why people become scientists.

Occupational Choice: A Personal Decision

Science is not for everyone. Most people lack not only the requisite skills for a scientific career but also the requisite interest. Science is prestigious perhaps because it demands not only talent but also persistence, dedication, and hard work, especially at the top. External rewards alone are unlikely to be enough to induce an ordinary person, even someone with scientific talent, to become a great scientist. The individual needs to be driven first and foremost by his or her own personal interest in science, and we do not know precisely how such interests are generated. While we cannot fully address the important question of interest generation here, we will devote a later chapter to the study of public opinions concerning science and scientists (Chapter 5) and another chapter to youths' aspirations to become scientists (Chapter 6).

Even excepting variation in taste, the vast literature on social stratification in sociology provides us with few clues about who will enter scientific occupations, as the literature is concerned mainly with vertical stratification—that is, how high a person's socioeconomic status is.[1] However, very different occupations (say, law and science) could be similar in terms of socioeconomic status. For individuals possessing different mixes of characteristics, entry into two occupations of similar social status may present quite different degrees of difficulty. In addition, people may differ significantly in what they value from work. For example, males are more likely to prefer jobs that give them high earnings and leadership opportunities, and females are more likely to prefer jobs that involve working with people and helping others or society.[2]

Due to a combination of many factors, such as ability, family environment, educational experience, personal taste, and chance, occupational preference is highly individualistic and thus highly heterogeneous across different persons. Furthermore, one's occupation, once chosen, can be changed only at a high, sometimes insurmountably high, cost in terms of lost prior investment in occupation-specific training, education, and experience. This is especially true for professionals, for whom changing one's occupation usually means discounting years of formal training and education. When this is the case, occupational choice is irreversible, or largely irreversible.[3]

These two factors—(1) variation across individuals in the difficulty of and attraction for entering different occupations and (2) largely irreversible occupational choices—leave open the possibility that young people will make consequential career decisions with limited information by selecting the future career trajectory that they believe maximizes their social statuses through the channel of mobility most appropriate to their known social characteristics and abilities. According to this reasoning, different individuals are likely to choose different channels of social mobility. At the macroscopic level, the large variation in ascribed and early-acquired characteristics, such as race, sex, social origin, talent, and school performance, will lead to a diversity of channels of social mobility.

Because individuals are not always able to make fully informed decisions, they base decisions on their perceptions of the advantages and disadvantages associated with various career options. These perceptions may be shaped by more advanced peers' career experiences, messages received from the media, or other types of influences. Whatever the source, these perceptions are, in turn, derivatives of the matrix of their social characteristics. As a result, individuals' social characteristics may affect their occupational choices through the perceived appropriateness of an occupation for someone with their traits or the perceived difficulty or rewards associated with a given occupation.

For people with certain characteristics, science is a preferred channel of social mobility. Given science's expressed norm of universalism, to be discussed below, science may serve as an attractive channel of social mobility for individuals who possess the educational credentials valued by science but come from disadvantaged family origins.

The Universalism Hypothesis

The universalism hypothesis, advanced by sociologist Robert K. Merton, postulates that in science universalistic (or impersonal) criteria are normally used to evaluate one's performance.[4] Through further elaboration, the universalism hypothesis argues that the norm of universalism should ensure that functionally irrelevant factors have no role in the evaluation process and the reward system.[5] The universalism hypothesis has two implications. First, scientific work is judged in terms of merit alone. Second, science recruits its members on the basis of talent and disregards functionally irrelevant factors such as race, gender, nationality, religion, and social origin.

The second implication of the universalism hypothesis compels us to be attentive to the key role of mediating factors. Our understanding of universalism does not preclude background factors from affecting factors that are functionally relevant to scientific careers. Rather, the question is, conditional on mediating factors, do particularistic factors that are functionally irrelevant by themselves exert any additional influence on entry into science? For example, we know that social background strongly affects educational attainment.[6] However, as long as the recruitment process into science is directly determined *only* by educational attainment, which is a universalistic (or impersonal) criterion, the process of recruitment to science can be said to be universalistic.[7] Accepting the universalism hypothesis as it relates to science does not mean that social inequality plays no role in determining access to scientific careers. Instead, it suggests that efforts to address this inequality should focus not on the decision to pursue science per se but on the process of educational attainment.

If universalism is indeed widely practiced in science, given the high social status of scientists, then at the individual level science can be viewed as an attractive channel of social mobility, because persons of disadvantaged social origins can effectively overcome their disadvantages through objectively measured criteria that all scientists accept. We are not aware of an argument that the extent to which the norm of universalism is practiced in American science has significantly declined over time. Thus, for persons who lack the physical, social, or cultural capital to enter other high-status occupations but possess scientific talents, science should remain an attractive occupation. However, if other high-status occupations have become more universalistic in applying similar criteria to those used in science, as has been proposed as a general trend in modern society,[8] science may lose some of its relative attractiveness. Competition for talent in other high-status professions such as medicine, law, and business management makes this hypothesis ever more plausible, given large increases in earnings in these occupations specifically—and associated with college education in general—since the 1980s.[9]

Rewards in Science

People are partly attracted to science by the rewards that it offers. These rewards are not purely economic: in addition to receiving higher than average earnings, scientists also enjoy high social status in modern society. Compared with the general population, they are highly educated, well paid, and well respected. In terms of occupational prestige, scientists are near the top, on par with Supreme Court justices.[10] Since 1977, the Harris Poll has included questions about the prestige of various occupations. This series of data shows some decline in scientists' prestige over the period, but the majority of Americans have consistently considered scientists as having "very great prestige," which is comparable to the prestige of medical doctors and surpassing that of lawyers, teachers, and bankers.[11] Recall our discussion in Chapter 1 that American science is Big Science, practiced by millions of workers.[12] The appreciable size of scientific occupations implies that a great many people can obtain high social status by becoming scientists. That is, many individuals may pursue scientific work for social status, economic well-being, and personal satisfaction, while at the same time meeting societal needs for scientific work.

We expect that most practitioners in the world of Big Science—scientific enterprise characterized by large teams of scientists and support staff working together on large-scale, big-budget research—today do not work mainly for high-profile recognition but simply to make a living in a highly skilled professional occupation that makes use of their scientific knowledge. In addition to money or recognition, other rewards such as social association, enjoyment of novelty and freedom, and power can also be motivating factors.

However, not everyone who wants to be a scientist can succeed at will. For the minority of scientists who seek positions as academic scientists, it takes long and arduous training (usually at least a PhD, often supplemented by postdoctoral work) just to earn the right to compete for these prized positions. Many who begin a science career later leave science. Hence, on their way up the career ladder from undergraduate education to a tenured position at a research university, many potential and novice scientists drop out either voluntarily or involuntarily. Like tournament players, only a small proportion of would-be scientists get to enjoy the optimal rewards of making original discoveries and gaining recognition for their contributions to science. Even among those fortunate few who are able to practice science as researchers, most just do what Thomas Kuhn calls "normal science" rather than making foundational discoveries as did Copernicus, Newton, Darwin, Mendel, and Einstein. It has long been recognized that inequalities in rewards—as well as productivity—are far greater in science than in society at large. For example, it is well documented that a small proportion of very productive scientists contribute a large fraction of the scientific literature,[13] leading Price to call the scientific contribution a process of "undemocracy."[14] This feature is epitomized in the tradition of eponymy in science—the practice of associating the name of a scientist to theories he or she has formulated, as with the Copernican model of the solar system, Newton's laws, the Gaussian distribution, or Einstein's relativity theory. Those scientists who make fundamental contributions to science may have their names permanently recorded in human history. Productive scientists have cumulative advantages because the resources for doing research are concentrated at institutions where the productive scientists are more likely to be located. Robert K. Merton aptly called these cumulative advantages of successful scientists the "Matthew effect."[15]

Although the high rewards and visibility of the scientific elite may seem unjust to those who equate social justice with equal outcomes, for others, the recognition of a few highly accomplished scientists is often couched in terms of the importance of their contributions to science, that is, a universalistic criterion that can be applied to all regardless of personal attributes. To the proponents of this view, because these scientific contributions are in the form of the public good—available to all and beneficial to all—the high status enjoyed by the scientific elite is socially justifiable.[16] As a result, elite scientists enjoy enormous prestige and power.[17] For the vast majority of rank-and-file scientists, even among the already small group of academic scientists, the chance of making a scientific breakthrough and/or winning significant recognition such as a major science award is small. Thus, the probability of success, especially high success, is low in science. Only those who believe that they possess special scientific talents, either through encouragement or self-discovery, are likely to aspire to high-flying careers despite the low odds.

Along the way, practical considerations of their likelihood of future success, either by themselves or by gatekeepers, may select individuals out of science, especially academic science. Hence, the lure of science in terms of making revolutionary scientific discoveries and altering the course of the human race, commonly associated with a handful of great scientists across history, does not match with reality for most scientists. Although the rewards for making a great scientific discovery are high, the chance of such a discovery is low, making academic science high-risk in terms of rewards. Most humans are risk-averse, preferring the "sure thing" to a lottery that gives them the same expected payout but a chance of leaving empty-handed.[18] Thus, if academic science is perceived as a lottery, individuals may choose instead to pursue careers with more certain rewards. This negative factor is perhaps now exacerbated by the lengthening of training (mostly postdoctoral), difficulties in finding suitable employment, and increasing difficulties in obtaining federal funding for research.[19]

Competition for a few highly prized status symbols has always been intense. Given that the scientific community worldwide has grown tremendously in recent decades, the odds of an individual winding up at the top of any scientific field have been getting smaller and smaller. Thus, the large variation in rewards may deter, more so than before, certain individuals from pursuing potentially rewarding but risky options in science if they treat the large uncertainty of outcome as a cost. If this is the case, we would expect that incentives to pursue careers in basic science, particularly in the academic sector, may have declined, while the attractiveness of careers in lower-risk alternatives, such as teaching rather than research, applied science, nonacademic research and development, or simply nonscience, will have increased. Thus, even if American science remains healthy in the production of science degrees and the number of employed scientists, the distribution of these positions may have shifted away from high-risk, high-reward positions focusing on basic research, potentially decreasing the productivity of American science.

Characteristics of Scientists

What types of individuals are most likely to develop scientific interests and aspirations? Starting more than a century ago with Francis Galton's *English Men of Science,*[20] the study of the characteristics of those most often recruited into science has attracted scholars from various disciplines: history, psychology, sociology, education, and policy studies.[21] The likelihood of becoming a scientist has been examined in relation to parental occupation, parental education, family income, religious origin, birth order, type of childhood, intelligence, marriage behavior, and many other sociological and psychological characteristics. These studies have mostly focused on eminent scientists, and their results are inconclusive.[22] An author portrayed—or rather, stereotyped—the average eminent scientist in the following passage in a 1968 book titled *Scientists in American Society:*

He was the first-born child of a middle class family, the son of a professional man. He is likely to have been a sickly child or to have lost a parent
at an early age. He has a very high IQ and in boyhood began to do a
great deal of reading. He tended to feel lonely and "different" and to be
shy and aloof from his classmates. He had only a moderate interest in girls
and did not begin dating them until college. He married late (at 27), has
two children and finds security in family life; his marriage is more stable
than the average.[23]

Other researchers studying scientists have often collected data in some arbi-
trary and nonrandom way. Some have obtained their sampling frame from a
national professional association, while others have simply based their studies
on "convenience" samples of scientists.[24] The pervasive use of nonrepresenta-
tive, nonrandom samples in studies of scientists highlights the unusual degree
of difficulty in conducting a formal statistical analysis of scientists.

Scientists are highly heterogeneous, despite a common job title. In fact, the
highly unequal distribution of rewards among scientists is a concrete indication of this heterogeneity. For this reason, any simple attempt to characterize
scientists is naive and flawed. When a researcher analyzes a group of scientists
who were not drawn randomly from a proper population of scientists, the result is highly dependent on who was included in the sample. In this study, we
mainly analyze nationally representative samples drawn randomly from known
populations so as to gain two advantages: first, we avoid arbitrariness of results
due to who happened to be included in a sample; second, we are able to characterize not only average attributes of scientists but also the extent to which
the attributes vary within a population of scientists.

How Scientists Are Made: Nature versus Nurture

Notwithstanding numerous empirical studies, few theoretical explanations
have been offered to account for the process of becoming a scientist. Two classic
theories meriting comment here are Francis Galton's inheritance thesis and
Robert K. Merton's Puritanism thesis. Inspired by his cousin Charles Darwin's
theory of evolution by natural selection, Galton was in constant search of statistical regularities pertaining to the intergenerational inheritance of intelligence
and, in the course of his search, discovered the phenomenon of regression and
formulated the concept of correlation.[25] During the "nature versus nurture"
debate that Galton himself initiated, he strongly maintained that inherited qualities were far more important than environment in determining the achievement
of individuals. As an additional proof of his general theory, Galton conducted a
survey of eminent scientists in England and hastily concluded that nature is

more important than nurture in producing eminent scientists after he observed that many of them came from families of high achievement.[26] The obvious shortcoming of Galton's analysis, judged by today's standards, is his inability to control for other factors, especially those relating to the environment. It is possible, for example, that children from successful families are more likely to be eminent scientists not because of inherited qualities but because of better environments in terms of more or better schooling, more leisure time, easier access to reading and experimental materials, and more frequent travel.

In short, how much scientific achievement is due to "nature" and how much is due to "nurture" is still an open question. Galton's position that everything is attributable to heredity can no longer be sustained in its original form. However, it is still useful to think about the often mixed, but structurally separable, effects of schooling and ability. Genotype ability cannot be directly measured. To further complicate matters, it is almost impossible to separate the effects of nature and nurture in a purely additive way, as the two sources of influence invariably affect individuals' outcomes interactively.

In contrast to Galton's emphasis on genetic inheritance, Merton's Puritanism thesis is essentially a "nurture" hypothesis offered to explain the rapid development of science and technology in seventeenth-century England.[27] His argument rests essentially on his Weberian interpretation of motivations of actors—scientists, in this case. Similar to Max Weber's analysis of the relationship between early capitalism and Protestantism,[28] Merton argued that the social values in seventeenth-century England were under the direct influence of a growing new religion, Puritanism, which called for more direct and effective ways of glorifying God. According to Merton, this Puritan calling transformed the mode of learning from scholasticism—that is, the study of Greek and Latin texts—into utilitarianism—greater focus on practical education—which in turn promoted the development of science and technology. To support his thesis, Merton traced biographies of eminent scientists and educational reformers and conducted statistical studies.

This is not the place to assess the rigor of Merton's analysis and the correctness of his thesis. It suffices to say that Merton's thesis simply cannot explain today's world. Merton has never pretended that it would. It is well known that in the United States, where there is a large degree of heterogeneity of religious and ethnic groups, Protestants are no longer overrepresented among scientists. Today, however, other demographic groups do stand out as being overrepresented in science, most notably Jews and Asians.[29] If we knew what accounts for the overrepresentation of these two groups, the answer might help us identify social factors that could be changed at the societal level to increase the talent pool for science. One popular explanation is cultural.[30]

However, the cultural explanation in its plain form has no analytical and explanatory value because it is circular: Jews and Asians have done very well in terms of educational attainment and scientific achievement because they value education and science. While we do not wish to reject the cultural explanation, we wish to reinterpret it within today's larger social and political contexts. Specifically, we propose that the universalistic norm in science may have encouraged the overrepresentation of Jews and Asians in science. While scientists today enjoy high social prestige and status, there is a general perception that anyone with scientific achievement is properly recognized, disregarding ascribed characteristics. In the face of minority status and potential discrimination, Jews and Asians may wish to find channels of social mobility that present fewer barriers but lead to high status. In the contemporary United States, science may be seen as an ideal occupation for those social groups, such as Jews and Asians, who are politically weak but academically successful.[31]

Summary

In a free-market society, the number of people who become scientists is fixed not by the policy decisions of those in power but by the collective decisions of individuals. The choice to become a scientist is a highly individualistic decision, but one that is also shaped by social context.

In this chapter, we discussed several different factors that seem likely to influence such decisions:

(1) Universalism: If youth with scientific talents in socially disadvantaged groups believe that they are more likely to be judged on the basis of merit alone in science than in other professional arenas, this may attract them to science, as science is viewed as a good channel of social mobility. That is, science is likely to be particularly appealing to those who lack the physical, social, or cultural capital to enter other high-status occupations but possess scientific talents.

(2) Rewards: Prospective scientists are attracted by both economic and other types of rewards, such as recognition, prestige, power, and pleasure in the work itself. These noneconomic rewards, however, are unequally distributed among scientists, especially in academia, being enjoyed very highly only by an elite group. Thus, students may feel that their chances of success are too low in science and opt for a career in which they can feel more certain that their efforts will be rewarded. This may encourage them to avoid riskier but potentially more rewarding options in science among those who become scientists or to forgo scientific careers altogether.

(3) Individual characteristics of scientists: The likelihood of becoming a scientist has been examined in relation to parental occupation, parental education, family income, religious origin, birth order, type of childhood, intelligence, marriage behavior, and many other characteristics. The relative importance of nature and nurture in scientific achievement has been repeatedly debated, but it has not been conclusively resolved. Previous studies focus mainly on select scientists and thus cannot be generalized to all scientists. Using data in random samples drawn from proper populations, we can characterize, statistically, attributes of scientists as well as the characteristics of individuals on their way to becoming scientists.

4

American Scientists: Who Are They?

He avoids social affairs and political activity, and religion plays no part in his life or his thinking. Better than any other interest or activity, scientific research seems to meet the inner needs of his nature.

—WALTER HIRSCH, 1968

In the last chapter, we looked at factors that may motivate individuals to choose science as a career. In this chapter, we will focus on the people who have already become scientists, both past and present. What are their demographic characteristics, and how have these characteristics changed over time? In answering these questions, we will be comparing American science not with science in other countries, as we did in Chapter 2, but with its own past as it evolved over the period between 1960 and the present. Within this historical framework, we will analyze changes in the size and demographic composition of the U.S. scientific labor force. We will also look at how scientists' rewards in the form of earnings have changed over the same period. Having finished this task, we will then spend the rest of this book exploring possible reasons for the observed trends in the American scientific labor force. Many different factors have been involved in creating and keeping this labor force, including the career choices of individuals, the rewards and social supports society offers the prospective scientist, the quality of science and math education at different levels, and the ease or difficulty of transitioning from school to work as a scientist. Thus, our approach, in this chapter, will be to start with our endpoint—the population of practicing scientists in America—and then go back and scrutinize the process of becoming a scientist and how this process and its results have changed over time.

Demographic Changes in the American Scientific Labor Force

Has the scientific labor force grown or shrunk in recent U.S. history? Using data from the 1-percent Public Use Microdata Samples (PUMS) of the 1960–2000 decennial U.S. censuses and the 2006–2008 American Community Survey (ACS) three-year estimates, we track trends in the scientific labor force, broken down by major fields of science and engineering, from 1960 through

2008. Unlike sample surveys, the decennial census attempts to enumerate and collect basic information about all individuals living in the United States at the time of the census. The very large sample sizes of the PUMS and ACS data are extremely helpful, as scientists represent only a small fraction of the general population, so a large sample is needed in order to include enough scientists to generate meaningful analyses. A drawback of the decennial census data is that data are collected only once every decade. The 2010 Census discontinued the practice of collecting information about work, as such information is now collected in the ACS. Therefore, we supplement our main analysis based on decennial census data with data from the ACS, which is administered annually by the Census Bureau. Each year, ACS collects information on a sample of American households, which generates estimates for the U.S. population as a whole. We use three-year estimates based on data collected between 2006 and 2008. For brevity, we refer to the ACS results as the results for 2007. The combination of the census data and the ACS data allows us to examine changes in characteristics of the U.S. scientific labor force in the second half of the twentieth century and into the first decade of the twenty-first century. More detailed information about our analysis using these data can be found in Appendix A.

We use the occupation-based definition of American scientists, considering them to be those employed in scientific occupations and holding at least a bachelor's degree. Military personnel and those who are not employed are excluded from our analysis. For the 1960 data, scientists are grouped into the major fields of biological science, engineering, mathematical science, and physical science. The data from 1970 onward allowed a separate categorization of computer science.[1] We present our results in Table 4.1.

Although always a small group, scientists constituted an increasing share of the labor force over this period, rising from 1.3 percent in 1960 to 3.3 percent in 2007 (shown in the first row of Table 4.1). According to these numbers, there is no evidence of a decline in the size of the scientific labor force during the second half of the twentieth century. On the contrary, the scientific labor force grew faster than the general labor force. However, some caution must be exercised in interpreting these trend data.

Because scientists are drawn almost exclusively from the ranks of persons with college degrees (and, in our definition, all scientists have college degrees), one may ask whether the scientific labor force has kept pace with the increase of the college-educated labor force that has resulted from the expansion of college education during this period. The answer is no. It seems that the growth of the scientific labor force lagged behind the growth of the nonscientific labor force for professional jobs that also required college education. We can see this by comparing the growth of the scientific labor force,

Table 4.1. Science, immigration, and education in the U.S. labor force, by decade (percent)

	1960	1970	1980	1990	2000	2006–2008
All sciences						
Scientists among employed	1.3	1.8	1.9	2.3	3.0	3.3
Employed with college degree or higher	11.1	14.7	22.4	25.7	29.6	32.9
Employed who are immigrants	6.0	5.7	8.0	8.8	12.5	17.0
Scientists among U.S.-born with college degree or higher	11.2	11.7	8.1	8.2	9.0	8.6
Scientists who are immigrants	7.2	10.5	12.6	14.5	23.7	27.5
Excluding computer science						
Scientists among employed	1.3	1.6	1.5	1.5	1.6	1.7
Scientists among U.S.-born with college degree or higher	11.2	10.2	6.4	5.4	4.9	4.6

Sources: U.S. Census 1960–2000; American Community Survey 2006–2008.

in the first row of data in Table 4.1, to the growth of the college-credentialed labor force, in the second row. While the scientific labor force increased by a factor of about 2.6 over the period, the share of college-educated workers in the overall labor force over this period increased by a factor of 3.0, from 11.1 percent in 1960 to 32.9 percent in 2007. This disparity is shown more clearly in row 4, which gives the percentage of scientists working among native-born Americans with at least a college degree. Here, we observe that the fraction employed in scientific occupations has fallen slightly, from 11.2 percent in 1960 to 8.6 percent in 2007.

Furthermore, much of the recent growth of scientific occupations appears to have been concentrated in computer science, rather than basic science, such as physical science and biological science. When computer science is excluded from the series, the fraction of the labor force engaged in scientific occupations grows much more modestly through the period, from 1.3 percent to 1.7 percent, and scientists as a share of college-educated native-born Americans fell more sharply from 11.1 percent to 4.6 percent. Of course, this large increase in the number of computer scientists was associated with the rapid expansion of information technology in the overall economy—"an industry that did not exist until the recent past," as characterized by the 2007 NAS report[2]—including growth in employment for computer specialists and technicians whose primary responsibilities may have included maintenance of computer networks and technical support to their end users. By examining trends when computer science is excluded, we do not mean to suggest that computer science is somehow not "real" science. Instead, by disaggregating, we highlight field-specific trends that are not always apparent when all scientific fields are lumped together. Table 4.1 presents the first of several instances in the book where breaking down results by field indicates a healthier trend for applied science than for basic science.

We now take a closer look at these numbers. Here are the major trends that our analysis reveals relating to the demographic composition of the American scientific labor force:

(1) A DRAMATIC INCREASE IN THE PROPORTION OF IMMIGRANTS RELATIVE TO NATIVE-BORN AMERICANS. A major trend in Table 4.1 relating to globalization is the growing representation of immigrant scientists. Over the period studied, the fraction of scientists who were immigrants rose sharply from 7.2 percent to 27.5 percent. We acknowledge, of course, that part of this increase reflects an increase in the overall immigrant population during this period. To account for this, we compare the percentage of immigrants among scientists with the percentage of immigrants in the overall labor force. The comparison shows that the increase of the former clearly exceeded that of the latter (from 6.0 percent in 1960 to 17.0 percent in 2007). That is to say, not only were immigrants more highly represented in scientific occupations than in the general labor force over the entire period, this overrepresentation increased throughout the period.

Immigrant scientists have always been a vital component of the American scientific labor force. By itself, a high representation of immigrants in scientific occupations need not be a cause for concern. However, future changes in immigration policy and increasing demand for and rewards to skilled labor in immigrants' countries of origins may diminish the ability of the United States to recruit immigrant scientists. A nation that fills more than one in four of its scientific positions with nonnative scientists may face considerable hardship if recruitment of foreign scientists becomes more difficult in the future.[3]

The rapid increase of foreign students in U.S. science and engineering programs and the parallel increase of immigrants among American scientists can be attributed to three causes.[4] First, less developed countries such as India, China, and South Korea can, and increasingly do, provide quality science education to their youth at low costs.[5] Second, science education is universal and thus transferable across national and cultural boundaries. Third, persons who grew up in less developed countries but received good science education have high incentives to immigrate to the United States for better work conditions and better economic rewards.

Furthermore, immigration and race/ethnicity are interrelated. Most immigrants today, once in the United States, belong to racial/ethnic minority groups. In addition, the high proportion of immigrant scientists has ramifications for racial inequality in the United States, mainly because science is among the most prestigious occupations in a modern society. Immigrants are sometimes perceived as taking up highly prized positions in science that otherwise would be available to native-born Americans, particularly members of racial/ethnic minority groups other than Asians.[6]

(2) A SHARP AND STEADY INCREASE IN THE REPRESENTATION OF WOMEN.[7]   As shown in Figure 4.1, while women comprised only 2.6 percent of the scientific labor force in 1960, they were about a quarter of employed scientists in 2007, compared with 46 percent of the general U.S. labor force in the same year.[8] That is, although still significantly underrepresented, women dramatically increased their representation in science in the second half of the twentieth century. This increase is at least partially explained by modest increases in women's representation in the U.S. labor force, from 31.5 percent in 1960 to 46.2 percent in 2007 and a larger increase in their

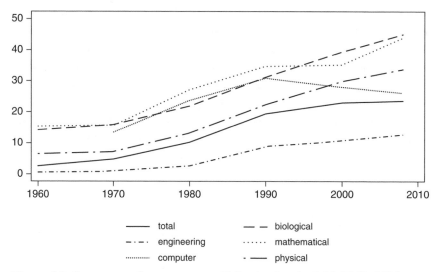

Figure 4.1. Percentage of women among U.S. scientists, by field, 1960–2008.
Sources: U.S. Census 1960–2000; American Community Survey 2006–2008.

representation among college graduates over the same period, rising from 26.5 percent in 1960 to 48.2 percent in 2007.[9] In 2007, women's representation in science varied significantly across scientific fields, with biological science and mathematical science leading, followed by physical science and computer science, and engineering trailing behind. Women already represented more than 10 percent of the labor force in biological science and mathematical science as early as 1960, and these numbers climbed to more than 40 percent in both fields in 2007.[10] By contrast, women were only 0.5 percent of engineers in 1960, but the percentage grew to 12.8 percent by 2007. Still, women were more underrepresented in engineering in 2007 than they were in either mathematical or biological science in 1960. Physical science and computer science occupy a middle ground, with women composing 26.1 percent of practitioners in computer science and 33.6 percent in physical science in 2007.

The steady increase of women in science is remarkable, especially given a common stereotype that women may have difficulty competing in science.[11] One of the major explanations proposed for the continuing underrepresentation of women in science is their difficulty in combining demanding scientific careers with family roles.[12] Men, in contrast, are often thought to benefit from marriage in their careers.[13] Has the differential impact of marriage on men and women scientists changed over time? In Figure 4.2, we show trends in the percentage of male and female scientists who are married. While male scientists were always more likely than female scientists to be married, the trend

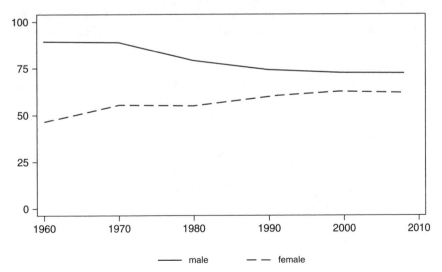

Figure 4.2. percentage married among U.S. scientists, by gender, 1960–2008.
Sources: U.S. Census 1960–2000; American Community Survey 2006–2008.

for men is downward, whereas female scientists were more likely to be married in 2007 than they were in 1960.

These trends must be interpreted in the context of a general trend away from marriage in American society. Among men, 89.4 percent of scientists were married in 1960, slightly higher than 87.8 percent for the general labor force.[14] Similarly, while only 72.2 percent of male scientists were married in 2007, this is still higher than the average of 64.6 percent for men in the general labor force. Thus, the declining trend in the marriage of male scientists reflects the general trend for the American population, rather than a trend unique to scientists. Male scientists are more likely to be married than the average male worker, and this gap has grown with time.

For women, we observe different results. First, the percentage of female scientists who were married first increased between 1960 and 1990 and then stabilized, around 60 percent, between 1990 and 2007. Second, the percentage of married women in the general labor force declined during the period, from 67.7 percent to 58.6 percent. As a result, female scientists were less likely to be married than female workers in general at the beginning of the period, but the gap closed by 2000. We do not detect a divergence in the gender gap over time: although it may be more difficult for women to combine married life with scientific work than it is for men, the differential difficulty does not appear to have worsened over time. If anything, the gap between female scientists and other working women declined over the period. Therefore, we find no evidence that science has lost ground to other occupations due to increasing difficulties of combining work and family.

(3) LARGE INCREASES IN AFRICAN AMERICAN AND ASIAN REPRESENTATION. Figure 4.3 documents trends in the representation of African Americans and Asians in scientific occupations. Figure 4.3a shows that the representation of African Americans in science increased sharply, from less than 1 percent in 1960 to 4.9 percent in 2007.[15] Still, African Americans remained underrepresented in scientific fields compared with their share of the American population during the period. African Americans represented approximately 9–11 percent of the U.S. labor force throughout the period, but their representation among college graduates rose from 4.0 percent in 1960 to 7.6 percent in 2007.[16] As with women, the increase in African Americans' representation in science is attributable partly to an increase in the fraction of African Americans who attained relevant educational training and partly to increased pursuit of science careers among African American college graduates.

In contrast with the trend for women, variation in African Americans' representation across fields shows instability over time. In 1960, African Americans were most highly represented in the biological sciences, but this

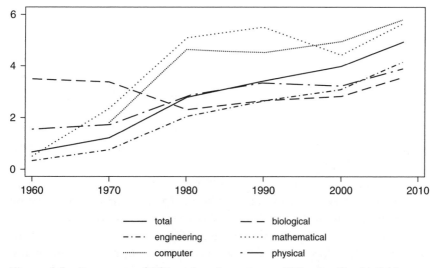

Figure 4.3a. Percentage of African Americans among U.S. scientists, by field, 1960–2008.
Sources: U.S. Census 1960–2000; American Community Survey 2006–2008.

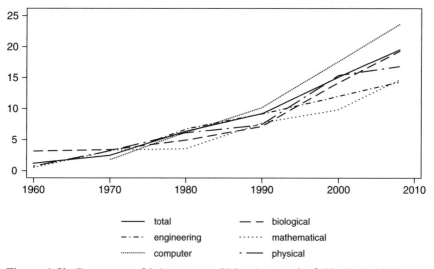

Figure 4.3b. Percentage of Asians among U.S. scientists, by field, 1960–2008.
Sources: U.S. Census 1960–2000; American Community Survey 2006–2008.

rate of representation hardly changed across the next forty-seven years, while African Americans made gains in other fields. As a result, by 2007, African Americans were most severely underrepresented in biological science. In 2007, the proportion of African Americans was highest in computer science (5.8 percent) and mathematical science (5.7 percent), followed by engineering (4.2 percent), physical science (3.9 percent), and biological science (3.6 percent).

Figure 4.3b shows the proportion of Asians among American scientists by field between 1960 and 2007. Asians accounted for only 1.2 percent of the U.S. science labor force in 1960 but 19.5 percent in 2007, when Asian representation was relatively high in all fields, varying from 14.5 percent in engineering to 23.6 percent in computer science. In fact, Asians' representation increased monotonically in all fields across the entire period. However, interpretation of these trends requires more care, because the fraction of the overall employed labor force identified as Asian also increased substantially, approximately sixfold during the period, from 0.6 percent in 1960 to 4.9 percent in 2007.[17] A similar trend occurs among college graduates: Asians were only 0.8 percent of the college-educated U.S. labor force in 1960 but 8.5 percent in 2007. Although Asians were always overrepresented among college graduates, their overrepresentation increased dramatically between 1960 and 1970 and then fell slightly between 1970 and 2007.[18] Still, these trends do not fully explain the rise in the representation of Asians among scientists:

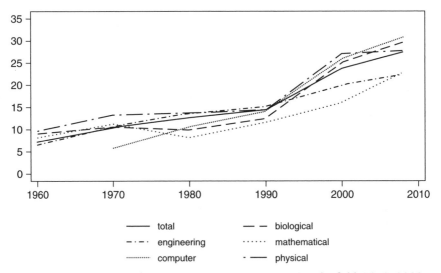

Figure 4.4. percentage of immigrants among U.S. scientists, by field, 1960–2008.
Sources: U.S. Census 1960–2000; American Community Survey 2006–2008.

the representation of Asian college graduates among the employed grew by a factor of about 11 compared with an increase by a factor of almost 16 in Asians' representation in the U.S. scientific labor force.

The rising share of scientists who are Asian is closely related to the secular increase in immigrants in the U.S. scientific labor force.[19] In Figure 4.4, we show a sharply increasing trend of the immigrants' representation among scientists during the period, from 7.2 percent in 1960 to 27.5 percent in 2007, a trend we showed in Table 4.1. In 1960, immigrants were only moderately overrepresented in the scientific occupations, constituting 6.0 percent of the total labor force. By 2007, immigrants' representation in science was 27.5 percent, about 60 percent higher than in the general labor force, which was 17.0 percent. A significant proportion of these immigrant scientists came from Asia, particularly India (16 percent of all U.S. scientists and engineers) and China (11 percent of all U.S. scientists and engineers). Immigrants are overrepresented in all scientific fields, and the variation in their representation across fields is not as pronounced as for women and African Americans. However, in 2007 their overrepresentation was somewhat higher in computer science, biological science, and physical science than in engineering and mathematical science.[20]

Taken together, these results suggest a changing portrait of the American scientific labor force. While native-born white men continue to dominate U.S. science, other groups—women, U.S.-born nonwhites, and immigrants—each comprise an increasing share of the overall population of American scientists. Besides the major immigration story, which we explained earlier, this trend is attributable to three distinct processes in recent U.S. history. The first is the marked increase in women's labor force participation.[21] Second, for African Americans and women, attainment of college education became more common with time, as access to higher education became more available after formal barriers were removed by the civil rights movement. Third, for all the nontraditional groups, including immigrants, participation in science among college graduates increased, suggesting that science became more open during this period, net of overall changes in the labor force and access to higher education.

The Earnings of American Scientists

In our earlier discussion of the factors relating to the attractiveness of scientific careers, we mentioned rewards. In Chapter 5, we will examine trends in the nonfinancial rewards of science, such as prestige. Now, we compare scientists' earnings with those of workers in other high-status professions and track trends in scientists' financial rewards. In examining the earnings data, we are driven by a specific hypothesis: the underpayment hypothesis. This hypothesis states that when faced with different career options, *ceteris paribus*, a person will choose the occupation that yields the highest expected monetary returns. At the societal level, when expected earnings of scientific occupations are relatively low compared with those of other high-skilled occupations, the overall interest in becoming a scientist decreases. While this reasoning seems straightforward, it actually has strong theoretical roots in several social science disciplines, including economics, sociology, and psychology.[22] In applying this reasoning, we ask whether over recent U.S. history the earnings of scientists have stagnated, while the earnings of other high-status occupational groups—law, medicine, or the social sciences, for example—have significantly increased. If the relative compensation of scientists declined, the decline may have channeled talent away from science into competing professions.

In this analysis, using the same census and ACS data described earlier, we use "basic" scientists—those in physical science, life science, and mathematical science—as our reference—and compare their earnings with those in other professions. We restrict our analysis to employed persons ages thirty-five to forty-five who are working full-time, year-round, and who report annual earnings of at least $5,000 in year 2000 dollars. We conduct our analysis separately for men and women and control for the individual's age and weekly hours worked. Because we wish to study scientists' earnings at different levels of education, we break up the comparisons separately to the bachelor's, master's, and doctoral degree levels. Our statistical methods are discussed in more detail in Appendix A. We would like to emphasize that due to data limitations, our statistical methods yield only approximate results that are confounded by both measurement error and sampling error. In particular, we should be skeptical of numbers that are based on a small number of cases (under 100); these are italicized in Table 4.2. Still, the results are informative in revealing broader trends in relative earnings of scientists and other professionals.

In Table 4.2, we present the core results of our analysis for men. We focus on men because the greater representation of men in science makes our estimates more precise for men than for women, particularly in the early years of the series, when female scientists were few in number. The analogous results for women are presented in Appendix Table D4.3. At both the bachelor's and the master's degree levels, we track trends over time in the relative earnings of engineers, computer scientists, nurses, teachers, and social scientists. While we consider computer scientists and engineers to be part of the overall scientific workforce, a more detailed analysis allows us to consider not only changes in the financial incentives for scientific as opposed to nonscientific

Table 4.2. Estimated ratios in earnings between professionals and scientists, by degree and decade (male workers only)

	1960	1970	1980	1990	2000	2006–2008
Bachelor's degree						
Scientists (bio., math, phys.)	1.00	1.00	1.00	1.00	1.00	1.00
Engineers	1.09	1.17	1.13	1.29	1.21	1.19
Computer scientists	N/A	1.10	1.09	1.19	1.26	1.22
Nurses	*0.81*	*1.12*	0.83	1.01	0.98	1.01
Teachers	0.68	0.70	0.69	0.77	0.71	0.67
Social scientists	*0.73*	*1.41*	1.17	1.07	*1.11*	1.18
Master's degree						
Scientists (bio., math, phys.)	1.00	1.00	1.00	1.00	1.00	1.00
Engineers	1.09	1.08	1.17	1.18	1.27	1.25
Computer scientists	N/A	1.11	1.09	1.07	1.31	1.27
Nurses	*0.68*	*0.53*	0.86	*0.98*	*1.07*	1.25
Teachers	0.68	0.71	0.77	0.78	0.77	0.74
Social scientists	*1.03*	1.08	*1.11*	1.00	*1.06*	1.22
Doctorate (PhD & professional)						
Scientists (bio., math, phys.)	1.00	1.00	1.00	1.00	1.00	1.00
Engineers	1.11	1.10	1.07	1.20	1.30	1.23
Computer scientists	N/A	1.06	0.93	*1.01*	1.37	1.22
Social scientists	*1.07*	1.12	1.02	0.97	1.19	1.13
Doctors	1.57	1.76	1.73	1.90	2.15	2.02
Lawyers	1.15	1.34	1.25	1.40	1.52	1.52

Sources: U.S. Census 1960–2000; American Community Survey 2006–2008.
Notes: Analysis is restricted to full-time, full-year workers.
Ratios are computed using scientists' earnings as the benchmark.
Estimates based on fewer than 100 cases are presented in italics.

careers but also changes in the rewards to subfields within science, particularly to applied as opposed to basic scientific labor. As discussed in Chapter 3, changing rewards within science may affect not only the attractiveness of science as a profession but the relative attractiveness of various subfields or job types within science. At the doctoral degree level, we present analogous comparisons for engineers, computer scientists, social scientists, medical doctors, and lawyers.

An entry in Table 4.2 represents the ratio in earnings of a profession (e.g., teachers) to scientists at a given level of education, everything else being equal. Among workers with bachelor's or master's degrees, we find that basic scientists generally earned less than engineers, computer scientists, and social scientists but more than teachers and (sometimes) nurses. Engineers with bachelor's degrees earned 9–29 percent more than basic scientists with the same education throughout the period. The highest premium for engineers was reached in 1990 (29 percent), but engineers in 2007 also had a higher premium (19 percent) than did engineers in 1960 (9 percent). For those with master's degrees, the premium for engineers has grown steadily, from 8–9 percent in 1960 and 1970, to 17–18 percent in 1980 and 1990, to 25–27 percent in 2000 and 2007. At the doctoral level, engineers enjoyed a higher level of premium between 1990 and 2007, in the range of 20–30 percent, than in the earlier period between 1960 and 1980 (7–11 percent). The premium also rose for computer scientists with a bachelor's degree, from 10 percent in 1970 to 22 percent in 2007. Computer scientists with a master's degree earned 11 percent more than similar basic scientists in 1970, but the premium then fell to less than 10 percent in 1980 and 1990, before rising to 31 percent in 2000 and 27 percent in 2007. At the doctoral level, computer scientists did not gain a large advantage until the end of the period, with a 37 percent premium in 2000 and a 22 percent premium in 2007. In summary, at all education levels, the average earnings of applied scientists in 2007 were significantly higher (by about one-fifth to one-fourth) than those of basic scientists, and the premium for applied science has generally increased since 1960.

What about the earnings of nonscientists? Social scientists with both bachelor's degrees and master's degrees generally outearn scientists of the same education level, but there are no clear trends in the amount of the premium. In 2007, social scientists earned between 13 percent (doctoral level) and 22 percent (master's level) more than similarly educated scientists.

For teachers, earnings were about two-thirds as high as scientists' at the bachelor's level and about three-fourths as high at the master's level, with no obvious trends. Nurses, however, showed gains compared with scientists at both the bachelor's and the master's levels. In 1960, nurses with bachelor's degrees earned 19 percent less than scientists, and the penalty for nurses with master's degrees was greater (at 32 percent). Beginning in 1990, how-ever, nurses at both education levels had no significant penalty compared with scientists, and at the master's level, nurses earned premiums compared with what scientists earned. Thus, scientists have lost significant ground in earnings compared with nurses but not with teachers.[23]

At the doctoral degree level, we find, unsurprisingly, that medical doctors and lawyers earned more than scientists throughout the period. For doctors, this premium grew over the period. Doctors earned 57 percent more than scientists in 1960 and almost 102 percent more in 2007. The premium for lawyers also rose substantially, from 15 percent in 1960 to 52 percent in 2007. Thus, scientists lost significant ground to both doctors and lawyers.

Taken together, the results show that scientists have not fared well in earnings in the past five decades. Compared with other elite professions we studied, the relative earnings of basic scientists have either lost ground or stagnated. More specifically, the earnings of basic scientists have declined relative to those of engineers, computer scientists, nurses, medical doctors, and lawyers and stayed similar in comparison to those of teachers and social scientists. Fed back to youth making occupational choices, this information could exert a negative influence on their aspirations to become scientists, particularly among youth with competing interests in fields that have seen rising relative earnings. Among those who pursue scientific careers, rising relative returns to applied science might draw youth away from basic science toward the more applied fields of engineering and computer sciences. Although absolute earnings of scientists and engineers increased over the period, the increases were so small as to constitute virtual stagnation in scientists' median earnings.[24] For male scientists and engineers with bachelor's degrees between 1960 and 2007, the annualized growth rate of real earnings was less than 0.5 percent, as it was for computer scientists with a bachelor's degree between 1970 and 2007. For male scientists with master's degrees or a PhD, annualized growth rates be-tween 1980 and 2007 were again no greater than 0.5 percent.[25] Over the same period, engineers and computer scientists with master's or doctoral degrees fared better, with annual growth rates between 0.8 percent and 1.5 percent. Our results on economic rewards suggest that perhaps basic science may have become less attractive to young people facing career choices during this period.

What our results do not suggest, however, is a market shortage of basic scientists. In a typical labor market, the intersection between supply and de-mand should determine how much employers are willing to pay for particular types of workers, that is, their earnings. If a shortage of scientists has devel-oped in the U.S. labor market over time, we should have observed a steady increase in earnings for scientists, as employers would need to pay higher salaries to compete for them. For basic scientists, stable or falling relative wages suggest, albeit indirectly, that there has been no shortage of scientists in the U.S. labor market. The increasing premium for applied as compared with basic scientists, however, may indicate a demand shift in the particular skills required of American scientists, particularly due to the expansion of the computer and technology sector.

One explanation for the declining trend of scientists' relative earnings is the displacement hypothesis. Related to the underpayment hypothesis, the displacement hypothesis proposes a particular reason why scientists are now paid less than before: the influx of immigrant scientists has increased the supply of scientists, resulting in scientists' lower earnings and thus lowering the monetary attractiveness of science occupations. An implicit assumption underlying the hypothesis is that scientific training and skills are more transportable across national and cultural boundaries than skills in many other professions, such as law and medicine. The logical argument of the displacement hypothesis is sound, as an increase in the supply of scientists should lower earnings of scientists. Indeed, there is some empirical support in the literature for the hypothesis.[26] The question, of course, is whether the magnitude of this displacement effect is sufficiently large to explain the observed trend in scientists' earnings. Short of experimental data, economic analyses in the past were bound to estimate this effect based on observational data and unverifiable assumptions. Because these assumptions are unlikely to hold true in reality, we do not believe that the displacement hypothesis has been proven. For this reason, we treat the idea that the immigration of foreign-born scientists is responsible for the declining trend in scientists' relative earnings merely as a hypothesis.

It is possible that overall trends in the earnings of scientists may mask divergent trends for demographic subgroups. Again using the census data, we found evidence of some convergence in the earnings of male and female scientists and white and nonwhite scientists.[27] To measure gender disparity, we use the female-to-male ratio in median earnings among full-time, full-year scientists ages thirty-five to forty-five. This measure of gender disparity in earnings is crude, of course, as it is confounded by other factors, such as detailed field of study, work activity, and labor supply. The inclusion of more detailed controls would increase the ratio toward parity in recent periods. The main interest lies in the change in this measure over time. By this measure, women earned 63 percent as much as similar male scientists in 1960, but their earnings grew to 86 percent of those of male scientists by 2007. This trend mirrors the decreasing gender gap in earnings in the general labor force, resulting in large part from overall increases during the period in women's commitment to labor market work.[28] The trend by race/ethnicity by the same measure is a bit surprising. The median earnings of African American scientists (as a single group) relative to those of whites rose from 80 percent in 1960 to 92 percent in 1970. For the next four decades, however, there has been little further progress in closing the race gap in earnings among scientists.

Summary

In this chapter, moving away from the global comparisons described in Chapter 2, we focused on the demographic characteristics and earnings of scientists within the United States, past and present. We began with what will become our endpoint, the population of working scientists, and looked at the ways both the demographic makeup and the earnings of this population changed during the period 1960–2007. Our findings, all relating to the period 1960–2007 in the United States, were the following:

(1) Scientists constituted an increasing share of the labor force over the period, but not enough to keep pace with the growth in the fraction of college-educated workers in the labor force.

(2) The fractions of women, immigrants, and racial minorities increased dramatically, to varying degrees from one scientific field to another.

(3) By the end of the period, women and African Americans were still underrepresented, while Asians had become increasingly overrepresented as the number of immigrant scientists increased.

(4) Relative to the earnings of those in other professions demanding similar education, the earnings of basic scientists generally declined over the same period, presumably making basic science less attractive to young people but also challenging the notion that a market shortage of basic scientists currently exists, as this would be expected to have raised earnings. Engineers and applied scientists fared somewhat better, with earnings increasing relative to basic scientists.

5

Public Attitudes toward Science

A scientifically literate society (not proficient, just literate) is essential to rational discourse and judgment in a millennium dominated by science and technology which to many people increasingly resembles sorcery.

—HOWARD SMITH, 2003

As discussed in Chapter 3, the process of becoming a scientist in any given society is the cumulative result of individual career decisions under the influence of the society at large. Most important to the well-being of science are the decisions of young people who are talented in science and math. But children and adolescents do not grow up in a vacuum. They are continually receiving messages from their surrounding cultures about what is and is not important, and these messages may have a great deal of influence over their future plans. If American culture devalues science, American youth may be ignorant of scientific discoveries and/or uninterested in pursuing scientific education or careers, a possibility we will explore in Chapter 6. In the present chapter, we draw on a variety of data sources concerning adult Americans' knowledge of and support for science in order to determine whether an erosion of either of these could be undermining pursuit of science by American youth. In particular, we make use of material compiled from the National Science Foundation's *Science and Engineering Indicators,* a biannual series of publications that presents quantitative data on various aspects of American science.[1]

In more concrete terms, we will examine some indicators of awareness of and support for science in American society and changes in these indicators in recent decades. This allows us to empirically assess changes in the social context in which prospective scientists are raised. If American youth grow up surrounded by adults who are relatively uninformed about or unsupportive of science, scientific training may seem unimportant or socially isolating to them. Of course, any indication of declining public support for science cannot be taken as direct evidence that this has been the *cause* of declining pursuit of science, as the two declines may share a common cause. A decline in scientific career choices, however, might be one particular manifestation of a larger public disenchantment with science. Examining general trends in knowledge of and public support for science can at least help us evaluate whether an individual's pursuit of science is discouraged by a broader disengagement from science.

Scientific Literacy

In a democratic society, public understanding of science is important and consequential. The more scientific knowledge members of a society have, the more enthusiastic they may be about supporting it.[2] Thus, we begin by considering the possibility of a decline in levels of scientific understanding among the general U.S. population in recent decades. Given the scope and significance of science and technology today, basic scientific knowledge is increasingly important to a public faced with the task of interpreting scientific information and using it to make decisions.[3] These decisions may be public in nature, such as deciding to endorse or oppose allocating public funds to the space program, stem cell research, alternative energy, climate change, or other avenues of scientific research. They may also be private, such as parental decisions concerning vaccination of children or consumers' decisions concerning genetically modified food.

The average level of scientific knowledge among adult Americans tells us little about whether young people are intellectually prepared to pursue scientific training, but the knowledge of those ordinary adults that young Americans come in contact with in their daily lives—parents and teachers who answer questions about the world, reporters in the media who report on scientific events, and voters who contribute to political decisions about science—may have important implications in shaping young people's attitudes toward science. Indeed, strong claims have been made that Americans' low level of scientific literacy presents a real threat to America's future.[4]

However, these claims are not fully supported by available data. We present trends in the scientific literacy of the American adult population over the period 1985 to 2006 in Figure 5.1. The overall picture that emerges from the data is one of low but constant levels of scientific literacy. The items are drawn from a set of ten mostly true/false questions that are designed to measure the respondent's knowledge of elementary scientific facts—often referred to as "scientific literacy."[5] As Figure 5.1 shows, considerable heterogeneity exists across items in the fraction of sampled respondents who answered correctly. When asked whether the center of the earth is very hot, whether the continents on which we live have been moving their location for millions of years and will continue to move in the future, and whether the earth goes around the sun or the sun goes around the earth, approximately three-quarters of the respondents answered correctly.

Three other questions, however, caused the respondents more difficulty, with less than half giving the correct answer in each year: whether electrons are smaller than atoms, whether the universe began with a big explosion, and whether "human beings, as we know them today, developed from earlier species of animals." Religious beliefs that conflict with evolution may partially explain why the majority did not answer "true" when asked whether humans developed from earlier species of animals. Another study with more detailed measurements has found that scientific literacy among adult Americans steadily increased between 1988 and 2007, with the exception of areas of science that conflict with the teachings of Christian biblical literalism.[6] However, the fact that Americans' scores are almost equally poor on a question that asks about the relative sizes of electrons and atoms, a question that appears unrelated to religious concerns, indicates that lack of scientific literacy is not limited to ideas challenged by religious groups in America.

Although we would like to see a higher level of scientific literacy among Americans in absolute terms, Americans' relative performance internationally is quite good. Whereas residents of the European Union (EU) perform similarly to Americans on tests of basic scientific knowledge, residents of China and Japan tend to score worse.[7] Overall, a higher proportion of American adults qualify as scientifically literate than adults in many other countries, a phenomenon largely attributable to the typical requirement, almost unique to the United States, that students in colleges and universities take at least

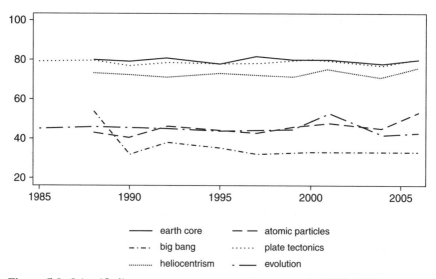

Figure 5.1. Scientific literacy: percentage answering correctly, 1985–2006.
Source: National Science Board (2008).

one year of science.[8] Thus, it seems to us an overstatement to call the United States "unscientific America" based on scientific literacy data.[9] Of course, belief in evolution presents an exception, as one study found that Americans lagged behind all other developed countries in the sample in acceptance of evolution, ranking above only Turkey.[10]

Public Interest in Science

The results shown in Figure 5.1 indicate that Americans possess moderate knowledge of basic scientific information of the kind typically acquired by adulthood. Furthermore, there is little evidence that this knowledge has declined in the last two decades. Such knowledge may be relevant to public issues relating to science. However, we have not yet asked how adult Americans learn about science. Do they acquire new knowledge beyond the classroom learning they experienced in their younger years? If Americans do not actively acquire new scientific knowledge, they are not informed of the latest scientific developments and are unlikely to be strong advocates for science. In other words, only knowledge about the latest scientific developments can help them form knowledgeable opinions on public policies concerning specific projects or programs in science. We know that following the Soviet Union's launching of Sputnik, for example, there was a great deal of public interest in and support for the government's investment in the space program.[11] Has public interest in science declined recently in America?

Overall, Americans appear to be at least as well informed about scientific facts relevant to either their personal lives or public debates as about abstract scientific knowledge. Surveys of the American public conducted in 2009 revealed that 91 percent of respondents knew that aspirin is recommended to prevent heart attacks, 61 percent knew that water was recently discovered on Mars, and 77 percent knew that underseas earthquakes can cause tsunamis.[12] These results show no evidence that contemporary Americans' scientific knowledge is limited to basic science of the kind learned in textbooks. But how does public interest in science today compare with that in earlier periods?

It is very difficult to accurately document changes in public interest in science. The difficulty lies in finding a consistent indicator with which we can track changes in public interest over past decades. One approach we have taken in this study is to evaluate trends in coverage of scientific topics by the news media. The indicator has some obvious weaknesses. First, it tells us only about media coverage, not media consumption. That is, we do not know if coverage was selected mainly by media elites or demanded by the public. Second, it is hard to know whether the meaning of media coverage has changed over time. For example, the growth of electronic media may mean that contemporary print media now reach a different audience and serve a different niche than in the pre-Internet era.[13] Still, our approach has the clear advantage of consistency over time in terms of format. Despite being an imperfect indicator, it provides some indication of the extent to which scientific information has been available in media and how this availability has changed over time.

We analyzed all cover articles published in *Newsweek* magazine from January 2, 1950, to June 25, 2007, comprising a total of 2,560 articles, and focused on the frequency with which these articles covered scientific news. We used a scoring system of 0–2 to denote how much scientific content a cover story contained.[14] We then computed a coverage index as an average for each year between 1950 and 2007. We present the results in Figure 5.2. Not surprisingly, the data are noisy. However, what is unmistakable in the figure is an overall upward, rather than downward, trend. In the first year, 1950, for example, the coverage index is 0.12, meaning that *Newsweek* published one full science cover article for every sixteen nonscience cover articles published. The index moved up during the period, indeed more than doubled to 0.35 for 2007, meaning that *Newsweek* published one full science cover article for every five nonscience cover articles published. This represented a significant increase in the coverage of scientific content in *Newsweek*.

A detailed look at the cover articles reveals that most of the increase in science coverage can be attributed to news in biological and medical sciences. Titles of medically or biologically focused articles included "Hormones" (Jan-

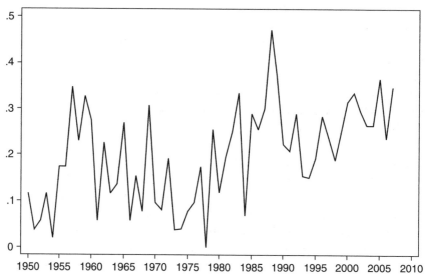

Figure 5.2. Science coverage index in *Newsweek*, 1950–2007.
Source: *Newsweek*, 1950–2007.

uary 12, 1987), "Where Health Begins" (September 27, 1999), "The Stem Cell Wars" (July 9, 2001), and "Genes and the Family" (February 6, 2006). Still, taken as a whole, there has been no decline in publication of science articles in *Newsweek*.

Because *Newsweek* articles were determined by media elites (i.e., magazine editors), the data may represent the editors' interests more than those of the public. To assess for potential bias in our data, we also conducted an analysis of books on the *New York Times* Best Seller List. We assume that popularity among ordinary readers caused a book to be on the *New York Times* Best Seller List. Again, what makes this data source attractive is that it allows us to go back all the way to the 1950s. We selected the months of March, June, September, and December as representative cross-sections of each year from 1950 to 2007, restricting our analysis to the top ten nonfiction titles.[15] Again, results from our analysis did not show any decline in interest in science. In contrast, there was an increase in interest in biological and medical science over the same period. Thus, the two different sources of data support the same conclusion.

This conclusion is consistent with survey data on Americans' own reports of interest in science. More than 80 percent of American adults say that they are interested in new scientific discoveries,[16] 35 percent claim that they enjoy keeping up with science news a lot, and two-thirds report regularly watching TV programs about science, such as *Nova* or those on the Discovery Channel.[17] In addition, public interest in science in the United States is also high relative to that in other countries. For example, Americans report more interest in science and technology news than do the residents of Europe, China, or South Korea.[18]

Nevertheless, between 1996 and 2008, the fraction of Americans who said they followed science and technology news closely fell both in absolute terms and relative to other types of news.[19] Some of this decline may have been due to a greater interest in medical news. Despite Americans reporting high levels of interest in scientific discoveries in general, reported interest in medical discoveries is even higher.[20] Medicine may even be beating nonmedical science at its own game. When asked whether various fields are "very scientific" or "pretty scientific," more than 90 percent of Americans believed that medicine, physics, and biology are each very or pretty scientific, but medicine received the most votes.[21]

Attitudes toward Science

Regardless of trends in Americans' scientific literacy or attention to scientific news, it is necessary to directly assess trends in the public's attitudes toward science. If the public's attitudes toward science have become less favorable in recent decades, this could lead to a reduction in attractiveness of science as a career option to youth. Recent studies reveal broad support for science and for public investment in scientific research. In 2009, nearly three-fourths of the general public agreed that government funding of basic scientific research pays off in the long run.[22] When asked about spending, 39 percent of the public reported that if they were in charge of the budget, they would increase spending on science. The same respondents, however, were even more likely to prefer increases in public spending on health care, education, and veterans' benefits.[23] The trends in support for science funding reveal a similarly mixed picture. Although support for increased spending on scientific research has risen over the last three decades, the rise in support for increased spending in other areas—including education, health care, environmental protection, and assistance to the poor—has been similar.[24] In relative terms, then, science is not a top funding priority and is not gaining ground compared with other recipients of federal funding.

Nevertheless, the public appears to maintain a very positive attitude toward science and scientists. According to the 2008 General Social Survey, an overwhelming 89 percent of the American public believed that science leads to more opportunities for the next generation.[25] In a 2009 Pew survey, 84 percent of the public agreed that science's impact on society has been mostly positive, and 70 percent believed that scientists contribute a lot to the well-being of society. Evaluations of medical doctors reached a similar level of approval, while only teachers and members of the military received more favorable assessments.[26] At the same time, Americans expressed some reservations about science, with a slight majority agreeing that science does not pay enough attention to the moral values of society and 47 percent agreeing that science makes our way of life change too fast.[27]

It is possible that Americans support science as an intellectual pursuit but not as a social institution. In this case, Americans might be reluctant to support funding increases for science if they believe that the funds will be misused by researchers. Figure 5.3 presents trends over the period 1973–2008 in the fraction of the population that report that they have "a great deal of confidence" in the leadership of a variety of social institutions. Over the entire period, Americans expressed more confidence in the leadership of the scientific community than in that of Congress, organized religion, or the press and more confidence in the leadership of the scientific community than in education leadership for all years since the early 1980s. However, Americans had more confidence in the leadership of medicine than in that of the scientific community over most of the period. Hence, no evidence exists either that Americans currently have low confidence in the scientific community or that confidence in the scientific community has declined over time. The fraction of Americans reporting high confidence in the leadership of the scientific community has remained fairly constant at around 40 percent since 1973, a period when confidence in medicine, organized religion, and the press fell substantially. Notably, while confidence in medicine was about 15 percentage points higher than confidence in science at the start of the period, the two had converged by the end of it.

Similarly, there is no evidence for a downturn in belief in the value of science. As shown in Figure 5.4, the fraction of Americans who agreed that the benefits of science outweigh the costs was stable (with some fluctuations) at around 70 percent between 1979 and 2008, and the fraction of the population endorsing federal funding for scientific research, even if it had no immediate benefits, went up from 79 percent in 1985 to 84 percent in 2008. But what about in earlier periods? A study published in 1981 on earlier trends found that the fraction of Americans who believed that science had changed life for the better fell from 83 percent to 71 percent between 1957 and 1976.[28] Combining information from the 1981 study with the more recent data in Figure 5.4, we observe that levels of support are somewhat lower in 2008 (68 percent) than in the late 1950s (83 percent), although there has been no significant decline since 1976 (71 percent).

In more recent decades, the prestige accorded to scientists relative to those practicing other occupations has declined somewhat, as shown in Figure 5.5. The fraction of Americans reporting that scientists have "very great prestige" declined from about 66 percent in 1977 to just under 56 percent in 2008. As a result, while slightly more Americans accorded "very great prestige" to

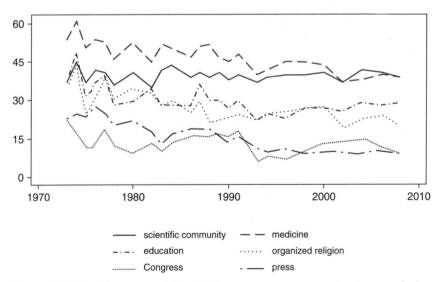

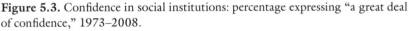

Figure 5.3. Confidence in social institutions: percentage expressing "a great deal of confidence," 1973–2008.
Source: National Science Board (2010).

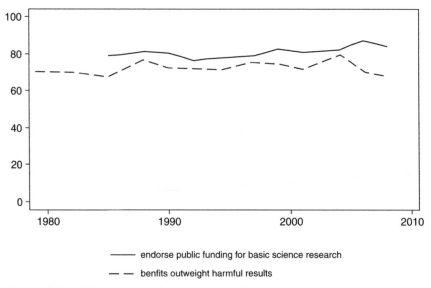

Figure 5.4. Public opinions of science, 1979–2008 (percent).
Source: National Science Board (2010).

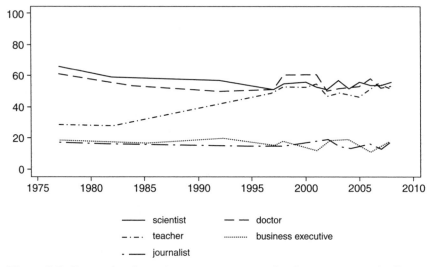

Figure 5.5. Occupational prestige: percentage reporting "very great prestige,"
1977–2008.
Source: National Science Board (2010).

scientists than to doctors at the beginning of the period and until the mid-1990s, the two lines have essentially converged by the end of the period.

Americans are not exceptional in their attitudes toward science. They are more favorable toward science than are Europeans and Japanese overall and even Chinese and South Koreans in some key respects. For example, compared with Americans, Chinese and South Koreans are less likely to say that they would be supportive of a son or daughter who wanted to be a scientist and more likely to say that science makes their way of life change too fast.[29] Among fifteen-year-old participants in the 2006 PISA test, U.S. students were slightly above the OECD average in their perceptions of both the general value of science (for such things as understanding the natural world, improving people's living conditions, and improving the economy) and the value of science to them personally (such as for understanding the world around them or for understanding how they relate to other people).[30]

Religion and Science

Among those concerned with American science, religion is sometimes viewed as a major barrier to the spread of modernism and scientific knowledge.[31] However, perceptions of the conflict between science and religion may be greater in public discourses than in actual situations experienced by individuals. While 55 percent of Americans believed that religion and science are often in conflict, only 36 percent reported that science sometimes conflicted with their *own* religious views.[32] Even among white evangelicals, a group that might be expected to have the highest risk of conflict between science and their religious beliefs, 48 percent did not report that their personal beliefs conflicted with science.[33]

We alluded earlier in the chapter to the proposition that religion may hamper Americans' acquisition of scientific knowledge, as Americans are less likely to agree with statements about evolution and the big bang than with most other basic scientific facts not in conflict with religious teachings. However, it is not clear that religious beliefs provide the full explanation. In the 2006 General Social Survey, which asked respondents the same true/false question about evolution, only 64 percent of respondents who identified themselves as having no religion asserted that the statement "Human beings, as we know them today, developed from earlier species of animals" was true. While this is considerably higher than the 52 percent of Catholics and 30 percent of Protestants who responded in the affirmative, it is still far less than 100 percent. This result illustrates that those answering "false" to this question are a heterogeneous group of individuals, as are those answering "true." Of those disagreeing with the evolution statement, some may be familiar with the concept of evolution but reject evolution in favor of the creation stories of their religion, while others—both the religious and the nonreligious—may simply be unfamiliar with evolution as a scientific concept. Similar results were obtained from a 2009 survey, which found that only 60 percent of those with no religious identification believed that humans had evolved over time due to natural processes (i.e., without the intervention of a supreme being).[34] While religious identification is clearly related to beliefs about the origins of human life, still a sizable minority of the nonreligious do not express belief in natural selection.

Similarly, the big bang theory also challenges some religious conceptions of the origins of the universe. Americans affiliated with religions that offer alternative explanations may be less likely than those with no religion to agree with the statement "The universe began with a huge explosion." Only 44 percent of those with no religion answered in the affirmative to this question, compared with 37 percent of Catholics and 25 percent of Protestants.[35] Again, these data suggest a role for religious beliefs as sources of rejection of scientific material, but they also reveal low levels of scientific literacy, for whatever reason, on these items among the nonreligious.

It is possible that advocacy by religious groups is responsible for Americans' low acceptance of the evolution and big bang theories.[36] For example, nonreligious individuals, though they have no personal reasons to reject natural selection or the big bang, may simply not have received scientific education on the topic because their education has been shaped by the religiosity of those around them. However, a study by the Pew Research Center in 2007 reported that more members of the American public believe that experts—scientists—agree on evolution than personally agree with evolution themselves.[37] This shows that many Americans are actually aware of the experts' opinions but choose to disagree. While we cannot establish what fraction of those who do not believe in natural selection do so because of a lack of familiarity with the subject, the results suggest that at least some individuals are not rejecting natural selection because they have never heard of the theory or are unaware that scientists support it.

Regardless of the relationship between individuals' religiosity and their scientific beliefs, we have not yet addressed the question of whether religious differences are also related to differences in Americans' support for science. The 2007 Pew survey found only small differences by religion in the beliefs that scientific developments have made society better, that the benefits of science outweigh the costs, that the federal government should support basic scientific research, and in reported happiness if one's son or daughter wanted to be a scientist.[38] In the 2009 Pew survey, 67 percent of those who professed some conflicts between science and their own religious beliefs also said that scientists have contributed a lot to society, compared with 72 percent among those without such conflicts.[39] Nevertheless, in the 2007 survey, the religious were more likely than the nonreligious to believe that science does not pay enough attention to the moral values of society and to agree with the assertion that we depend too much on science and not enough on faith.[40] In summary, religious Americans may be characterized as almost equally supportive of science's accomplishments compared with the nonreligious but with somewhat greater reservations.

Summary

Do Americans support the work of scientists? Our answer is overwhelmingly yes, though not without some qualifications. Here are our findings:

(1) Compared with members of other countries, Americans are neither notably ignorant in scientific matters, nor less interested in them, nor less supportive of science.

(2) Americans have remained interested in scientific news, although their interest has shifted in recent decades away from basic science toward subjects that concern them in their daily lives, such as health.

(3) Americans believe that science has helped society and have great confidence in the scientific community.

(4) Americans support public funding for scientific research and bestow high prestige on scientists, though scientists' prestige has declined slightly in recent years.

(5) While Americans' religious views may sometimes conflict with scientific theories, public perceptions of the frequency with which this occurs may be overstated, and religious faith does not appear to diminish support for the work of scientists in any significant way.

Finally, we note that medicine plays an ambivalent role in public interest and support for science. On the one hand, concern for medical decision making may motivate Americans to pay attention to scientific discoveries with medical implications. On the other hand, Americans' interest in health news may substitute for, rather than complement, their interest in other types of scientific news. While Americans hold scientists in high esteem and have confidence in the leadership of the scientific community, doctors rival scientists in these domains. Americans are more interested in medical news than in news on basic science and even consider medicine to be more scientific than the basic science fields. A medical career requires intensive scientific training and may be perceived as a close substitute for a scientific career. The high prestige and high pay of medical doctors, as well as considerable public interest in medicine, may serve to encourage students with scientific talent to pursue medical careers rather than careers in the basic sciences.

6

Does Science Appeal to Students?

No one wants to be such a scientist or to marry him.

—MARGARET MEAD AND RHODA MÉTRAUX, 1957

In the previous chapter, we found little evidence that Americans' attitudes toward science and scientists have become less favorable over recent decades. Scientists have continued to receive considerable support and prestige from the American public. This does not reveal, however, whether science remains an attractive career path for young Americans. American youth may believe that science is a worthwhile pursuit for our society in general, while finding it unattractive as a career choice for themselves.

Who, then, would pursue a science career? We propose that prospective scientists must meet at least three requirements. First, they must believe that they can receive appropriate training to *qualify* to do scientific work. Second, they must consider science a *desirable* profession. Third, they must *prefer* science to other potentially attractive career options. In the preceding chapters, we discussed some factors that may affect the third requirement: while the nonfinancial rewards in science as compared with those in other professions have remained high, in the form of support for science and the prestige of scientists, financial rewards in basic science relative to those in other high-status professions have declined. In this chapter, we present evidence pertaining to the first two requirements and then examine the total impact of these social factors on the fraction of youth who intend to pursue scientific training and careers.

The scientific pipeline starts with science education. We begin by examining trends in adolescents' training in science, as well as in their stated expectations regarding pursuing a college degree in a scientific field. For most scientific jobs, a bachelor's degree in a scientific field is the minimum prerequisite. However, the path to a scientific career often starts prior to college. High school students who aspire to scientific careers may begin to take advanced courses in math or science in preparation for later pursuing a bachelor's degree in science. In this way, young students' expected college majors may have attitudinal and behavioral implications in high school that ultimately affect their likelihood of completing bachelor's degrees in science. Therefore, an analysis of high school students' aspirations is a useful precursor to an analysis, in Chapter 7, of the attainment of bachelor's degrees in science.

Precollege Education in Science

In the 2008 General Social Survey, 70 percent of adult Americans reported being dissatisfied with the quality of math and science education in American schools.[1] This concern is reflected in the alarm, discussed earlier in Chapter 2, over American students' performances on international math and science tests. However, there is little evidence that American students are receiving worse education in math and science today than in past decades. If anything, students may now be receiving better education in these subjects, especially math. The National Assessment of Educational Progress (NAEP), a nationally representative assessment of American school-aged children, has tracked trends in students' knowledge of various academic subjects since the 1970s. From 1973 to 2008, the average math scores of both nine-year-old and thirteen-year-old students increased significantly, while the scores of seventeen-year-olds remained flat. Furthermore, gains have been larger for African American and Hispanic students than for white students during this period, reducing the racial gap in math achievement.[2]

An assessment of the trend in American students' average achievement in math is, however, not fully informative about trends in the potential pool of scientists, who are disproportionally drawn from the top tier of the academic distribution. Has there been a decrease in the performance of American students with the highest levels of academic achievement in science-related subjects? Again, this does not appear to be the case. The math scores of students at the 90th percentile rose significantly between 1978 and 2008 for nine- and thirteen-year-olds, while remaining flat for seventeen-year-olds.[3]

There are also some signs of improvement among high-achieving high school students. Among seventeen-year-olds, 19 percent had taken precalculus or calculus in 2008, compared with only 6 percent in 1978.[4] Furthermore, the number of students both taking and passing Advanced Placement (AP) exams in math and science subjects increased rapidly between 1997 and 2008.[5] In summary, there is no evidence that today's American schoolchildren are less prepared than their counterparts of three decades ago to enter advanced training in scientific fields.

Self-Concept Regarding Math and Science Abilities

Students are more likely to pursue fields in which they perceive that they are talented. This relationship is mediated in part by academic achievement. That is, students' perceptions of their own abilities in math and science are associated with their academic achievements in these subjects, which, in turn, are associated with their interest in scientific careers.[6] However, influences of math self-concept on course-taking behavior remain even after accounting for the mediating role of academic performance.[7]

Unfortunately, we did not find good data with which to assess trends in math self-concept across time for American youth. Thus, we cannot tell whether self-concept in math has been falling in America, nor can we determine whether race and gender differences in mathematics self-concept, net of performance differences, have narrowed over time. However, if self-concept in mathematics differs across demographic groups at a given time, net of performance and achievement, and if these differences contribute to group differences in interest in scientific careers, this suggests the potential importance of policies targeting students' self-concepts: the pool of talented students pursuing training in scientific fields will increase if the self-concepts of all American youth regarding math and science improve.

Most existing research on differences in mathematics self-concept has focused on gender differences. Regarding race, there is little evidence that, controlling for achievement test scores, racial minorities have lower self-concepts in math than do whites. In fact, after controlling for both test scores and grades, whites have lower self-concepts in math than do African Americans, Hispanics, or Asians.[8] In contrast, large gender differences in math self-concept remain, in favor of men, net of gender differences in achievement.[9] Since, as previously discussed, math self-concept is predictive of persistence in math and science fields, at least one reason for women's lower levels of pursuit of scientific training may be their lower perceptions of their own mathematical abilities, regardless of their actual performances.

Images of Scientists

Students who are talented in scientific subjects and confident in their abilities may still refrain from pursuing scientific studies. Not only may falling wages for scientists relative to those for other professionals discourage youth from scientific careers, but youth may also avoid science if they perceive that scientists are very different from themselves or that the lifestyles of scientists are not compatible with their own desires. Surveys of American youth that ask them to describe their perceptions of scientists, either in words or by drawing a scientist, provide insight into American students' perceptions of scientists and scientists' lives. The earliest such study was performed by the celebrated anthropologist Margaret Mead and her coauthor, Rhoda Métraux, and revealed that American high school students have generally positive but highly stereotyped views of scientists, frequently describing scientists as middle-aged or elderly men wearing lab coats and glasses, working alone, and surrounded by scientific equipment.[10] Scientists were perceived as being intelligent, well-trained, hard-working, careful, and patient, working for the good of humankind rather than for personal gain. At the same time, scientists were seen as engaging in monotonous work that often takes years to yield results, for low pay and little recognition. Furthermore, scientists were perceived as being personally boring, engaging little in social life, neglecting their families, and having few interests outside their work.[11]

More recent studies also confirm this finding. American students appear to respect those who become scientists, while at the same time finding scientific work and the scientist dull, suggesting that most students do not believe that scientific work would be personally rewarding. In fact, when tenth and twelfth graders are asked to list their reasons for not intending to major in science, the most common reason given is simply that science is boring.[12] Assessments conducted by NAEP in 1977 and 1982 revealed that less than 30 percent of seventeen-year-olds found their science classes fun.[13] This occurred despite the fact that more than three-quarters of the same students believed that science was generally useful.[14]

These attitudes and stereotyped images of scientists and their work have persisted among American youth.[15] For example, pictorial images of scientists drawn by students continue predominantly to depict white men, despite increasing representation of women and racial minorities in scientific occupations.[16] While these stereotyped images of scientists may indeed have discouraged American youth from pursuing scientific careers, there is little evidence that stereotypes have worsened over recent decades. If there is any change, the images of scientists have become less stereotypical and more positive over time.[17] Furthermore, similar stereotyped images of scientists appear to be present in some other countries, including both Western (Canada, Ireland, United Kingdom, Australia) and non-Western countries (Taiwan, Korea).[18]

Thus, we are led to conclude that poor perceptions of scientists should not have contributed much to a decline, either absolute or relative, in pursuit of science by American youth. However, providing American students with more diverse and appealing images of scientists may encourage them to pursue science. In particular, students may be more likely to consider science a viable career path when they are presented with images of scientists that are similar to their conceptions of themselves. It is possible that the high-profile images of scientist-entrepreneurs, such as Bill Gates and Larry Page, inspire youth to be interested in science. As of 2010, seven of America's twenty richest individuals were self-made billionaires of the technology boom.[19] Certainly, the potential for this kind of financial reward, as well as the associated fame and prestige, may motivate some individuals to pursue technological careers. It is worth noting, however, that the software tycoons are not purely (perhaps not primarily) scientists as they are entrepreneurs and CEOs. With the possible exception of these few highly successful entrepreneurs in the computer industry, however, images of scientists may remain largely negatively stereotyped.

Belonging to the World of Science

One implication of the stereotyped images of scientists is that American students may find a science career unsuitable for them. In other words, they may be concerned that they would not fit into the world of science. The question of belonging may be particularly salient for women and racial minorities who have historically been underrepresented in scientific fields.[20] A psychological study demonstrated this fact by randomly assigning a group of students into one of two experimental conditions: to list either two or eight friends who had traits that would enable them to fit in well in computer science. Listing two friends was an easier task than listing eight friends. Students were then asked questions designed to assess their own perceived level of fit with computer science. While white students' perceived fit was not affected by the number of friends they were asked to list, African American students who were asked to list eight friends had a lower perceived fit in computer science than those who were asked to list two friends.[21] In other words, when African American students were led to believe that they did not know many other people who were suited to computer science, they were more likely to believe that they did not belong in computer science either.

Among young women, the perception that science is a poor match for persons who are interested in helping others may discourage their pursuit of science. In a study of high schoolers' career interests, females were more likely than males to talk about wanting jobs that would allow them to help and care for people, and some females explicitly mentioned intending to avoid science as a major because it was too impersonal and did not relate to people.[22] Even among females who intended to major in a scientific field, scientific training was often seen as a necessary step on the way to a nonscientific career that is perceived to provide help to others, such as a health profession.[23]

Expectations Concerning Scientific Majors

Having explored the major factors that may affect students' desires to pursue a scientific career, we now devote the remainder of the chapter to presenting observed trends in American youths' aspirations regarding scientific fields of study or occupation. In Figure 6.1, we show trends in college freshmen's interest in science as a future occupation or major. The data come from the Cooperative Institutional Research Program (CIRP) Freshman Survey, which has collected information from students at a variety of bachelor's degree–granting institutions since 1966. Because the sampling procedure for institutions changed in 1971, we present trends beginning in 1971. Every year, more than 200 institutions and over 100,000 freshman respondents participate in the CIRP Freshman Survey. Despite the large sample size, observed trends in interest in science are somewhat noisy, perhaps indicating that short-term fluctuations in scientific interest respond to current events or current labor market conditions.[24] Figure 6.1a shows trends in interest in science as an occupation. If we ignore short-term fluctuations, it appears that interest in employment as a scientific researcher fell slightly in the late 1970s, followed by a period of relative stability. The percent of respondents indicating that they were likely to become scientific researchers first declined steadily from 3.3 percent in 1971 to 1.7 percent in 1982 and then fluctuated between 1.7 percent and 2.4 percent in subsequent years. Although this decline in the 1970s is small in absolute terms, it is large in relative terms. At least some of this decline might be accounted for by rising interest in becoming a computer programmer over the same period, increasing from 1.1 percent in 1971 to a high of 8.3 percent in 1982. Interest in computer programming waxed and waned over the rest of the period, falling to 2.0 percent in 1992, then rising to 5.2 percent in 2000, and then falling to 1.7 percent in 2006. The two peaks of interest in computer science—first in the early 1980s and then at about 2000—correspond roughly to the expansion of the personal computer and the dot-com boom, respectively. The computer science fluctuations thus provide one example of interest in fields responding to changing market conditions. Similarly, interest in engineering rose from a low point of 5.3 percent in 1971 to a high point of 11.1 percent in 1982, followed by a period of decline, to 6.3 percent in 2006. Thus, it appears that interest in computer science and engineering positions peaked in the early 1980s, while interest in scientific research fell throughout the period.

For comparison, we also show trends in the percentage of students expressing interest in becoming doctors, lawyers, and teachers. Interest in teaching declined rapidly in the 1970s, from 17.3 percent in 1971 to a low of 5.4 percent in 1982, at about the time that interest in engineering and computer science was peaking, but then recovered somewhat over the rest of the period, fluctuating between 9 percent and 11 percent since 1990. By contrast, interest in medicine and law were both fairly stable until about 1990, at about 5 percent for each, after which interest in medicine rose somewhat to a high of 7.6 percent in 1995 while interest in law fell to a low of 3.8 percent in 1996. Figure 6.1b presents trends in interest in particular scientific majors over the same period and tells a similar story. As before, we see peaks in interest in

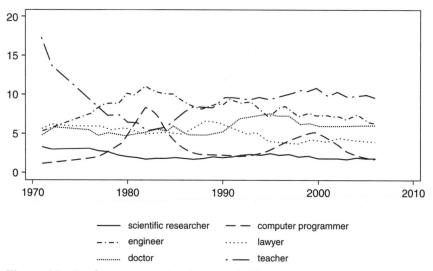

Figure 6.1a. Freshman interest in science: probable occupation, 1971–2006 (percent).
Source: Pryor et al. (2007).

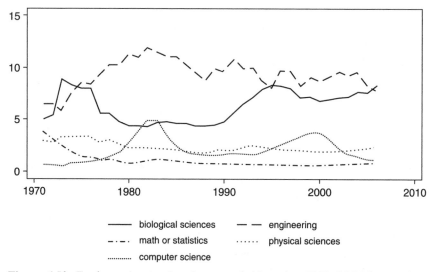

Figure 6.1b. Freshman interest in science: probable major, 1971–2006 (percent).
Source: Pryor et al. (2007).

computer science and engineering in the early 1980s, with a second peak in the late 1990s for computer science. At its peak in 1982, 11.9 percent of freshmen expected to major in engineering, compared with 6.5 percent in 1971 and 7.8 percent in 2006. Interest in computer science also peaked in 1982–1983, at 4.9 percent, compared with 0.7 percent in 1971 and 1.1 percent in 2006. After a drop after the mid-1970s, interest in biological science increased beginning in about 1990. By 2006, 8.3 percent of freshmen reported biological science as their probable major, compared with a low point of 4.3 percent in 1982. This may be related to the increased interest in medicine over the same period, as majoring in biological science is often considered a useful precursor to medical training. Interest in physical science declined slightly over the period, from 3.4–3.5 percent in 1973–1976 to under 2 percent in the years around 2000, before it increased in more recent years. Interest in mathematical science plummeted sharply in the 1970s, from 3.7 percent in 1971 to about 1 percent in the years around 1980, then to less than 1 percent in every year following 1987.

In the remainder of this chapter, we will present results from our own analysis. The results presented here should be read in tandem with the analogous discussion of trends in the receipt of scientific bachelor's degrees in Chapter 7. While only a minority of students expecting to receive a bachelor's degree in science ultimately receive one, scientific aspiration is a powerful predictor of completing a bachelor's degree in science for both men and women.[25] Comparing trends in expectation and attainment allows us to distinguish trends in interest in science from trends in actually achieving scientific training. For both analyses, we make use of longitudinal data on three cohorts of youth collected by the National Center for Education Statistics (NCES). Each cohort of school-aged youth was followed through high school graduation and for at least eight subsequent years, with several follow-up interviews during this period. The NCES data sets provide information about the high school graduating classes of 1972, 1982, and 1992, respectively: the National Longitudinal Study of the Class of 1972 (NLS-72), High School and Beyond (HS&B), and the National Education Longitudinal Study of 1988 (NELS).[26] By comparing the experiences of these three cohorts of students, we are able to track changes over time in American youth's aspiration for and attainment of college-level scientific training. Appendix B provides more detailed information about the analysis with the NCES data sets.

Our examination of students' educational expectations decomposes expectation formation into two sequential steps: (1) the expectation of receiving a bachelor's degree in any field, and (2) among those students expecting to receive a bachelor's degree, the expectation that the degree will be in a scientific field.[27] Implicitly, we assume that the expectation of college graduation is a precursor to expectations regarding a particular field of study at the college level.

In each of the three data sets, the survey asked students in their twelfth-grade year a question about their expected level of educational attainment and field of study in college. In Table 6.1, we show trends across cohorts in the share of men and women who expected to receive a bachelor's degree and, among those who expected to do so, trends in the share of students of each sex who expected that they would major in a scientific field.[28] For both sexes, the fraction of students who expected to receive a bachelor's degree was highest in the 1992 cohort, with 65.7 percent of young men and 69.6 percent of young women expecting to receive a bachelor's degree. For the 1972 cohort, the corresponding percentages were 56.6 percent and 46.0 percent.

Among those expecting to receive a bachelor's degree, there is some evidence of declining interest in science. In the 1992 cohort, only 27.5 percent of men and 10.5 percent of women expecting to receive a bachelor's degree anticipated a scientific major, as compared with 35.8 percent and 14.3 percent, respectively, for the 1972 cohort. However, these aggregate numbers mask differences across subfields. For both men and women, the fraction of students who expected to receive a bachelor's degree in the life sciences, conditional on expecting a bachelor's degree, declined: from 14.5 percent in the 1972 cohort to 3.8 percent in the 1992 cohort for men, from 9.4 percent to 3.7 percent for women. Expectations of receiving a degree in engineering increased for both sexes: from 11.3 percent in 1972 to 16.3 percent in 1992 for men and from 0.2 percent to 2.8 percent for women.

In addition to trends for all high school seniors, we also examine changes in interest in science among those students who are best prepared for college-level scientific study. If science has become less attractive to those students who are most likely to make contributions to its advancement, this alone may pose a problem for American science, even if the overall number of students pursuing science remains stable. Thus, we focus on students who scored in the top quartile of a math achievement test administered by each of the three surveys. For students of both sexes, there is some evidence of a decline in interest in science among the top quartile of math achievers who expect to receive a bachelor's degree. In the 1992 cohort, among men, 38.4 percent of these students expected to major in a scientific field, compared with 47.5 percent in 1972 and 53.0 percent in 1982. For women, 17.5 percent of top achievers expecting a bachelor's degree anticipated majoring in science in 1992, compared with 24.6 percent in 1972 and 20.7 percent in 1982. Thus, interest in science among high-achieving students appears to

Table 6.1. Student aspirations for bachelor's degree, by gender and cohort (percent)

	Male			Female		
	1972 cohort (NLS-72)	1982 cohort (HS&B)	1992 cohort (NELS)	1972 cohort (NLS-72)	1982 cohort (HS&B)	1992 cohort (NELS)
Expecting bachelor's degree	56.6	44.6	65.7	46.0	45.6	69.6
Among top 25% in math achievement	86.3	80.3	92.9	82.6	81.9	95.2
Expecting science major given bachelor's degree	35.8	41.4	27.5	14.3	13.5	10.5
Among top 25% in math achievement	47.5	53.0	38.4	24.6	20.7	17.5
Expectations of science subfields given bachelor's degree						
Physical science	5.0	3.7	2.6	1.2	1.7	1.6
Life science	14.5	3.7	3.8	9.4	2.2	3.7
Mathematical science	3.4	0.9	0.8	2.7	1.1	1.0
Computer science	1.6	10.9	4.1	0.8	5.2	1.4
Engineering	11.3	22.3	16.3	0.2	3.3	2.8

Sources: NLS 1972; HS&B 1980; NELS 1988 (see Appendix B).
Note: The sum of percentages across the science subfields does not exactly equal the percentage of science majors due to double-majoring and rounding.

have followed much the same trend as interest among the general student population.

In order to explore the determinants of expecting to receive a scientific bachelor's degree, we apply our two-step conceptualization and model the probability of (1) expecting to receive a bachelor's degree and (2) expecting to receive a degree in science, conditional on expecting to receive a bachelor's degree. Each of these outcomes is modeled separately by sex and cohort, using race, family income, maternal education, and family structure as predictors. In some models, we also include measures of the child's math achievement and of the difference between his or her math and verbal achievement scores. For brevity, we refer throughout to the models that exclude achievement score measures as the "demographic" (simple) models and to models that add the achievement measures as the "score" (full) models. In the demographic models, we are able to observe aggregate differences in educational expectations by traits that measure students' social origins. We expect students' academic achievements to be predictive of their expected levels of educational attainment as well as interests in science, net of other demographic factors. Furthermore, we conjecture that a student's achievement in math relative to reading, for comparative advantage reasons, is also positively predictive of interest in science.

We compare the results of the demographic models to those of the score models to learn about the sources of differences in students' educational expectations. The results from the demographic model reveal aggregate differences in educational expectations by students' social origins. Comparing the results between the demographic models and the score models, we can observe the extent to which these differences by social origin are mediated by students' academic performances. If significant demographic differences in interest in science remain after adjusting for academic performance, it is possible that perceptions of science and scientists contribute to making science more appealing to members of some social groups than to members of others. If, conversely, demographic differences in expectations of pursuit of science are greatly reduced or eliminated after adjusting for students' academic achievements, this suggests that demographic disparities in students' scientific training are the main reason why the groups differ in interest in science. Throughout, we focus on the results for the most recent cohort, although differences across cohorts are generally small.[29]

As expected, students' math achievement scores are significantly and positively associated with their likelihood of expecting to receive a bachelor's degree, for both sexes in all cohorts. Furthermore, among students who expect to receive a bachelor's degree, those with higher math achievement are also more likely to expect to study science. However, contrary to our expectations, students' math achievements relative to their reading achievements are fairly weakly associated with their likelihood of expecting to pursue science. It is possible that high school students' senses of their relative abilities are only weakly developed: strong students may perform well in high school courses in all areas, providing them with little information either about their relative talents or about how to pursue careers according to them.

After we control for academic achievement in the statistical models, results for demographic groups differ by gender. Among men, there are significant racial differences in expecting to receive a college degree in both the demographic and the score models.[30] In both models, net of other covariates, white men have the lowest likelihood of expecting to receive a bachelor's degree. Since African American and Hispanic men have lower average achievement scores than do white and Asian men, the higher expectations of African American and Hispanic men become even more pronounced once achievement is controlled for in the score model. Among those who expect to receive a bachelor's degree, however, there are no significant racial differences in either the demographic model or the score model.

For women, the story is reversed: there are no significant racial differences in the expectation of receiving a bachelor's degree in either model, while there are significant racial differences in expectations of receiving a scientific degree among those who expect a bachelor's degree. These differences are driven by the much higher interest in scientific degrees among African American women expecting bachelor's degrees in both the demographic and the score models, the effect being more pronounced in the score model, again due to the lower average achievement scores of African Americans.

Finally, we examine whether students from more privileged backgrounds—those with higher family incomes and those whose mothers received more education—are more likely to express interest in science degrees. In the demographic models, more advantaged students of both sexes are significantly more likely to expect to receive a bachelor's degree. These differences are reduced but not eliminated in the score models. Thus, parental advantage increases students' educational expectations both indirectly, through its effects on academic achievement, and directly, net of achievement.

Among students who expect to receive a bachelor's degree, the relationship between family advantage and interest in science is much less clear. Among male students, after adjusting for academic achievement, there are no significant differences by maternal education, although males from more financially advantaged families are significantly less likely to profess interest in a science major. Among female students in both models, neither maternal education nor family income is significantly associated with an interest in a science major among those expecting a bachelor's degree.[31]

In general, our results are consistent with universalism in science. First, social origins exert a much larger effect on students' expected level of educational attainment than on the expected interest in science among those who expect to receive college degrees. Second, to the extent that demographic differences exist in expectations of scientific education, they favor those from disadvantaged social origins. If science is perceived as universalistic by youth, it should be more attractive to youth from disadvantaged social origins. Among young women expecting to complete a bachelor's degree, African Americans are most likely to expect to receive a science degree. Among young men, those from families with fewer financial resources are more likely to expect to receive a science degree.

Summary

Compared with the glowing perception of science and scientists presented in the previous chapter, the results presented in this chapter show that American youth exhibit an ambivalent view toward science. Recall our three prerequisites for prospective scientists: perception of adequate training, desirability of a scientific career, and preference for a scientific career over other available career options. What have we learned about each of these factors as they pertain to contemporary American youth?

(1) There is no evidence of decline in the mathematics training provided to American students. Average scores of elementary and middle school students have increased in recent decades, and more high school students than before are now taking advanced math courses and receiving college credit for their high school mathematics work.

(2) Results of studies that ask students to draw scientists, to write about scientists, or to report their own reasons for not wanting to be scientists show a consistent picture: while scientists are generally esteemed by young people as hard-working members of a prestigious and important institution, many students perceive science and scientists as dull. Moreover, these perceptions are fairly stable across time.

(3) There is some evidence of decline in interest in science among American youth, although the decline is moderate and sensitive to the particular indicator that is used and which specific field of study is considered.

(4) Trends for high-achieving students generally mirror those of the general population. Students who performed well in math, both male and female, were more likely to expect to pursue scientific education than the average student in all periods.

In addition to trends across time, we are also concerned with differences in scientific expectations among different subgroups of high school students:

(1) Young women are much less likely to expect to pursue science than are young men. Women's lower mathematics self-concept, net of mathematics achievement, may partially explain this result.

(2) Consistent with universalism, there are few differences in students' expectations regarding scientific study by race and socioeconomic status among those who expect a college education. The differences we do see tend to favor less advantaged students—young men from low-income households and African American young women. Most of the disparities by social background lie in expected level of educational attainment rather than in the interest in science net of expected level of educational attainment.

7

Attainment of Science Degrees

Genius without education is like silver in the mine.
—BENJAMIN FRANKLIN, 1750

As discussed in the Introduction, two alternative definitions of "scientist" are commonly used in studies of scientists: (1) the occupation-based or demand-based definition; (2) the education-based or supply-based definition.[1] Whereas the occupation-based definition identifies as scientists the incumbents of scientific occupations, the education-based definition considers individuals with or working toward science degrees to be potential scientists.

In Chapter 4, we relied on the occupation-based definition when we examined scientists' demographic characteristics and labor force outcomes. However, we are keenly aware of two obvious shortcomings of the occupation-based approach. First, all persons who receive science education make up the potential supply of the scientific labor force. Although for a variety of both market and personal reasons only some of them actually end up working in scientific occupations, the number of persons with science education is still a useful indicator of the strength of a nation's scientific workforce. Second, we already know that today's process of becoming a scientist involves elaborate and largely continuous science education. A narrow focus on only those persons already in scientific occupations would miss the important causal process involved in creating scientists—science education—and would not allow us to predict whether there is likely to be a shortage of scientists in the future.

We have already documented trends in a variety of aspects relating to the well-being of American science, including, in Chapter 4, the demographic characteristics of American scientists; in Chapter 5, attitudes toward science by the broader American public; and, in Chapter 6, expectations of attaining science degrees. We now examine the attainment of American science education at the college and graduate levels, which is critical for determining whether American science will become vulnerable in the future.

Science always requires education, but education does not have to be scientific. Earlier in the book, we considered science education and career interests at the pre-college level. In this chapter, we shift our attention to the attainment of science degrees at the bachelor's, master's, and doctoral levels. We begin with an examination of broad trends in attaining science degrees by both de-

90

gree level and field of study. We then document the trends by two core demographic variables: gender and immigration status. Next, we report results from a detailed analysis of changes in the attainment of science degrees at the bachelor's level for three cohorts of American high school seniors with longitudinal data—the 1972 cohort, the 1982 cohort, and the 1992 cohort. We also consider some of the reasons college students cite for switching majors away from science. Finally, we analyze data from several sources on the transition from science education at the bachelor's level to graduate degree programs.

Broad Trends in Science Degrees

To examine trends in scientific training, we analyzed federal statistics on science degrees awarded in U.S. institutions of higher education. We draw data on the production of science degrees over the past four decades from the Integrated Postsecondary Education Data System (IPEDS) Completions Survey, conducted by the NCES, through its online tabulation system.[2] The data, described in more detail in Appendix C, are sufficiently detailed for us to construct three time series of scientific degree production at the bachelor's, master's, and doctoral levels between 1966 and 2008.

We first present the overall numbers of science/engineering (S/E) degrees by year and degree level in Figure 7.1.[3] The trend in bachelor's degree production, shown in Figure 7.1a, is an increase in the number of degrees awarded,

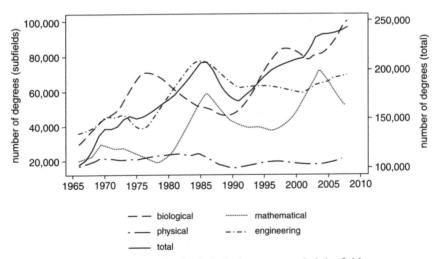

Figure 7.1a. Science/engineering bachelor's degrees awarded, by field, 1966–2008.
Source: WebCASPAR Integrated Science and Engineering Resource Data System (2010).

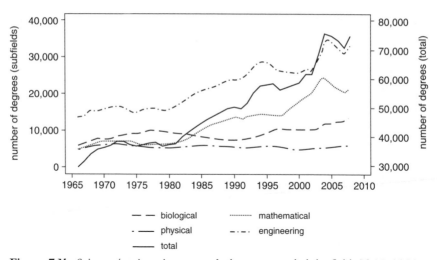

Figure 7.1b. Science/engineering master's degrees awarded, by field, 1966–2008.
Source: WebCASPAR Integrated Science and Engineering Resource Data System (2010).

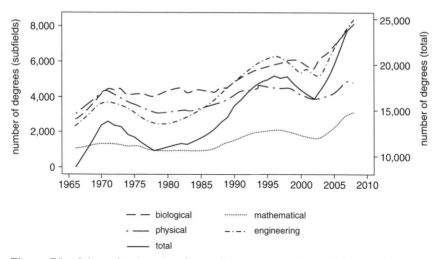

Figure 7.1c. Science/engineering doctoral degrees awarded, by field, 1966–2008.
Source: WebCASPAR Integrated Science and Engineering Resource Data System (2010).

although there is a slump between the mid-1980s and mid-1990s. In raw numbers, the number of bachelor's science degrees awarded rose from 102,983 in 1966 to 242,684 in 2008. This amount of growth is equivalent to an annualized growth rate of 2.1 percent, far short of both the growth rate in the pre-1960 era (4.7 percent) and our estimated growth rate of the scientific labor force during the post-1960 period (3.9 percent).[4] The gap in growth between the scientific labor force and the production of science bachelor's degrees may be accounted for by immigrants working as scientists in the United States who had completed bachelor's degrees abroad prior to immigration.

The growth in the number of bachelor's S/E degrees is most pronounced in the life sciences, with an annualized growth rate of 2.9 percent. As a result, by 2008, 100,480 degrees, or 41.4 percent of the total number of bachelor's degrees in science, were given in biological science. Growth in the mathematical sciences was the second most rapid, at the annualized rate of 2.2 percent.[5] Engineering degrees also experienced moderate growth, at an annualized rate of 1.6 percent. Growth in the physical sciences was very little, at an annualized rate of 0.6 percent. To put those growth rate numbers into perspective, we note that the U.S. labor force grew at about 1.7 percent annually between 1960 and 2007, so that an annual growth rate below 1.7 percent meant a relative decline in reference to the overall U.S. labor force.[6] In other words, between 1960 and 2007, the number of individuals with bachelor's degrees in the biological and mathematical sciences more than kept pace with growth in the overall labor force, whereas degrees in engineering and the physical sciences did not.

The trends at the graduate levels are similar. The production of both master's and doctoral science degrees, shown in Figures 7.1b and 7.1c, respectively, increased during this period, with some short-term fluctuations. The total number of master's degrees in science grew steadily from 29,545 in 1966 to 74,668 in 2008, at an annualized rate of 2.2 percent. The overall increase in the production of science doctoral degrees was higher than at the other two lower levels during the same period, from 8,829 in 1966 to 24,493 in 2008, at an annualized rate of 2.5 percent. Recalling our reference of 1.7 percent annual growth in the labor force, the increase in the production of both master's and doctoral degrees in scientific fields more than kept pace with the growth of the labor force. Of course, much of the growth at the doctoral level was driven by foreign students, a topic to be discussed in more depth below. The production of science PhDs accelerated between 2002 and 2008, growing at 7.1 percent annually, for a total increase of 51 percent in just six years.[7] As with bachelor's degrees, the number of master's and doctoral degrees granted in each field rose over the period, but the growth was always fastest for biological science, followed by mathematical science, engineering, and then physical science.

If the production of science degrees is taken as a measure of the well-being of science, the trends presented in Figure 7.1 should not cause great alarm. American universities have continued to produce graduates with degrees in science at all levels of education. Furthermore, the production of science degrees has generally increased over time, although the magnitude of the increase is modest. This increasing trend in potential scientists is particularly impressive in recent years at the doctoral level. But was the upward trend mainly driven by international students from abroad? Was it driven by the influx of women into science? To answer these questions, we now turn to a more detailed analysis of the trends by gender and immigration status.

Trends in Science Degrees by Gender and Immigration Status

In Figure 7.2, we present trends in the composition of the recipients of S/E degrees by gender and immigration status. As we expected, the representation of women graduates in scientific fields, shown in Figure 7.2a, grew dramatically at all levels over the period.[8] In 1966, women constituted 16.2 percent of recipients of bachelor's degrees in science, 9.6 percent of those receiving master's degree in science, and just 5.8 percent of those receiving doctoral degrees in science. In 2008, forty-two years later, women constituted, in science, 38.5 percent of bachelor's degree recipients, 32.6 percent of master's degrees recipients, and 33.4 percent of those receiving doctoral degrees. Although women continued to be somewhat underrepresented, the growth in women's representation has occurred at all levels and has been more pronounced for more advanced degrees. In a supplementary analysis of the trends by field of study, we found the trend of women's increasing representation in all four major fields (i.e., biological science, physical science, mathematical science, and engineering) and at all three levels of education (i.e., bachelor's, master's, and doctoral degrees), except for bachelor's degrees in mathematics, which showed a decline.[9] Women's increasing representation in biological science was particularly dramatic, women in 2008 being in the majority at the bachelor's level (58.2 percent) and the master's level (57.4 percent) and reaching parity at the doctoral level (49.5 percent).

Women's increasing representation among science degree recipients accounts for part of the overall increase in the production of science degrees. When we restricted the analysis of the trends to men, the increase in science degrees was much less (about half the overall increase) but still substantial. At the bachelor's degree level, the increase in degree production between 1966 and 2008 was 73 percent among men, compared with 136 percent overall; at the master's level, the contrast was 88 percent versus 153 percent; at the doctoral level, it was 96 percent versus 177 percent. That is to say, half of the total increase in the production of S/E degrees was among women, although they comprised less than half of the degree recipients.

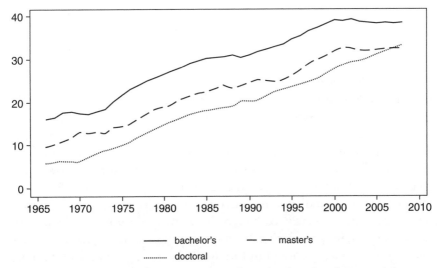

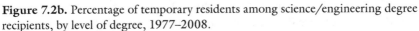

Figure 7.2a. Percentage of women among science/engineering degree recipients, by level of degree, 1966–2008.
Source: WebCASPAR Integrated Science and Engineering Resource Data System (2010).

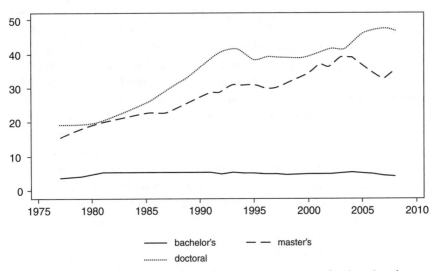

Figure 7.2b. Percentage of temporary residents among science/engineering degree recipients, by level of degree, 1977–2008.
Source: WebCASPAR Integrated Science and Engineering Resource Data System (2010).

In the debate on U.S. science policy, much attention has been paid to the large proportion of foreign students receiving doctoral education in the United States.[10] Why should these trends be of interest in the debate concerning the well-being of American science? Three issues have been raised. First, does the large flow of foreign students, particularly those from Asian countries, crowd out native-born Americans in doctoral science programs?[11] Second, does the training of foreign students at the doctoral level, particularly at elite institutions, weaken the global position of American science if highly trained foreign students return to their home countries following graduation?[12] Third, if foreign students stay in the United States after completing their doctoral degrees, do they contribute toward lowering the earnings of scientists overall and thus discouraging native-born Americans from pursuing science careers?[13] Although our analyses do not directly address these concerns, we provide documentation of the underlying claim that science degrees from American institutions are increasingly awarded to immigrants, rather than to native-born Americans.

Figure 7.2b shows trends in the share of science degree recipients who are temporary U.S. residents. Temporary residence is the best available measure of foreign students in government statistics on degree recipients, and it is available only from 1977 onward. Among science degree recipients at the bachelor's level, the fraction of temporary residents has remained fairly constant since 1977, never exceeding 5.5 percent. The fraction of temporary residents was much higher for advanced degrees throughout the period. Among scientific master's degree recipients, it rose rather steadily from 15.6 percent in 1977 to 35.3 percent in 2008. Among scientific doctoral recipients, the share of temporary residents grew more rapidly, from 19.3 percent in 1977 to 46.7 percent in 2008. By the end of this period, about half of all doctoral S/E degrees were awarded to foreign students. While temporary residents have always been more highly represented among recipients of more advanced science degrees, trends in the last thirty years have exacerbated this gradient, with representation among scientific doctoral degrees rising fastest of all.

Does the increasing participation of foreign students in American science education at the master's and doctoral levels account for the growth of science degree production that we documented earlier in Figure 7.1? To answer this question, we reanalyzed data restricting the sample to degree recipients who were U.S. citizens or permanent residents. As expected, the growth at advanced levels of science degree recipients who were either U.S. citizens or permanent residents was much lower than the overall growth during the 1977–2008 period. At the master's level, the total growth rate was reduced from 92 percent to 47 percent after we removed foreign students from the calculation. Similarly, at the doctoral level, the growth rate was reduced from 117 percent to a mere 44 percent.[14] For comparison, we note that the growth rate excluding temporary residents was 43 percent at the bachelor's level. Thus, when temporary residents are excluded, the growth rates were fairly comparable across the three levels of educational attainment: only the increased representation of temporary residents at higher degree levels has led to the more rapid increase in the production of advanced degrees. We conclude that, as with women, the influx of foreign students accounted for a large share of the increases in degree production at the master's and doctoral levels.

There were differences across fields. The share of foreign students was highest among recipients in engineering and mathematical science. In 2008, for example, foreign students accounted for 40 percent of master's and 61 percent of doctoral degrees in engineering and for 43 percent of master's and 56 percent of doctoral degrees in mathematical science.

What is the combined effect of these trends at the doctoral level? To explore this question, we use data from the Survey of Earned Doctorates (SED) and Survey of Doctorate Recipients (SDR) and compare the composition of degree recipients in 1966 and 2006.[15] Overall, the number of doctoral degrees awarded in science grew by 248 percent in total, or 2.3 percent per year. However, for male U.S. citizens and permanent residents, the total increase has been only 4.1 percent, or 0.1 percent per year. By contrast, the number of doctoral degrees in science awarded to female U.S. citizens or permanent residents has grown at a rate of 6.1 percent per year. Among temporary residents, the analogous rates are 9.1 percent per year for women and 4.4 percent per year for men. In summary, the number of men who are American citizens or permanent residents obtaining science degrees has indeed stagnated over the past four decades.

As a result of these disparate growth rates, the composition of the recipients of scientific doctoral degrees has changed considerably. In 1966, 79.5 percent of scientific doctoral recipients were male U.S. citizens or permanent residents, while only 14.7 percent were male temporary residents, 4.8 percent were female U.S. citizens or permanent residents, and 1.0 percent were female temporary residents. By 2006, about two-thirds of all recipients of scientific doctoral degrees from American institutions were men, and they were equally split between U.S. citizens or permanent residents and temporary residents. Among women recipients, U.S. citizens and permanent residents still outnumbered temporary residents, 20.3 percent to 13.2 percent, although both groups had increased their representation relative to that of men. In 1966, women were 5.7 percent of U.S. citizens and permanent residents receiving scientific doctoral degrees, but by 2006 their share had increased to 38.0 percent. Women's representation among temporary residents has been somewhat smaller, from 6.4 percent to 28.4 percent.

Two-Step Conceptualization for the Attainment of Science Degrees

What social processes underlie the trends that we have examined in Figures 7.1 and 7.2? More specifically, has the likelihood that an American youth with certain characteristics would complete a science degree changed over time? Put another way, is there good evidence for a popular view that there has been "a growing aversion of America's top students—especially the native-born white males who once formed the backbone of the nation's research and technical community—to enter scientific careers?"[16] If American science is losing talent to alternative career options in the United States, in the long run, such a decline in interest could negatively affect the well-being of American science.

To study how individuals' characteristics affect the likelihood of completing a science degree, we break down the process of degree attainment into two sequential steps: (1) attainment of a degree regardless of field and (2) attainment of college-level science education given a degree. Thus, our strategy here is parallel to our two-step decomposition of students' educational expectations in Chapter 6. As before, this decomposition strategy presumes that the choice to pursue science education is a decision secondary to that of completing a degree. Although this assumption may not hold true in all situations, it greatly simplifies our conceptualization of the research question, as it allows us to separate out explanatory factors that predict the choice of science as a potential field of interest from explanatory factors that affect the attainment of a degree per se, regardless of field.

Social Determinants of Science Careers

Underlying the preceding two-step conceptualization of the process of attaining science education is the recognition that social mechanisms determining the choice of science versus nonscience careers may be different from those determining the attainment of a college education. From past research, we already know very well that education, especially college education, has become an increasingly important social determinant of life chances and lifestyles in America.[17] There are two primary explanations for college attendance. First, college-going may be motivated by the rational expectation that a college education would yield material returns, such as higher earnings, higher social status, better health, and more civil engagement.[18] Second, college-going may be the result of a cultural norm based on one's own socioeconomic or class background.[19] Either way, past research has extensively studied the level of educational attainment, but it has paid little attention to field specialization at a given level of education that is closely linked to an occupation. In the language of the above two-step decomposition, the literature has been mainly concerned with the first step, that is, attainment of education, but not with the second step, that is, choice of science given a particular level of education. We have good reasons to believe that the social determinants affecting the two steps are different.

This difference stems from a distinctive feature of science—universalism—that was discussed in Chapter 3. The universalist norm requires not only that scientific results be verifiable but also that "the acceptance or rejection of claims entering the lists of science is not to depend on the personal or social attributes of their protagonist; his race, nationality, religion, class, and personal qualities are as such irrelevant."[20] As discussed in Chapter 3, the combination of the high social status of science and the norm of universalism may make science a good channel of social mobility for individuals from less advantaged backgrounds.[21] That is, persons who lack physical, social, or cultural capital important for other high-status occupations but who possess scientific talents can prove their "worthiness" to be scientists as judged by universalistic criteria. This is essentially an argument about comparative advantage: those individuals who perform well in mathematical and scientific subjects and prefer objective criteria for evaluation may view science careers as an advantageous means to achieving high social status and high earnings. In contrast, the process for educational attainment is different, being affected strongly by one's social background conditional on academic performance. That is to say, although both science and higher education are associated with high social status, the influence of social advantage on science education, conditional on higher education, may be entirely mediated by academic performance, in a way that is not true with respect to the level of educational attainment in general. We already saw some evidence consistent with this conjecture in the analysis of students' educational expectations presented in Chapter 6. In this chapter, we test whether the hypothesis also holds true for educational attainment.

Social Changes Affecting Science Education

The focus of the current study is on trends. Is there any reason to believe that the pursuit of science education by American youth has declined? What social changes might have affected their likelihood of attaining science education? We propose two factors. First, in an increasingly globalized environment, American youth face increasing pressures from the competition of immigrants, including those from countries with lower overall standards of living but stronger mathematical and scientific education, such as China, as discussed in Chapter 2. The comparative advantage argument posits that

American youth may face a disadvantage relative to foreign competition. If students are forward looking and wish to pursue training in fields in which they perceive a high likelihood of secure employment in the future, this relative disadvantage may steer more Americans now than previously toward fields other than math and science.

Second, as we saw in Chapter 6, even among students who do not share this concern, science may have lost some of its appeal. In addition to the nonfinancial factors we discussed, a decline in scientists' earnings or prestige relative to those in other professions should also make science careers less attractive. This reasoning is consistent with the classic discrete choice model in economics, rational choice theory in sociology, and social learning theory in psychology.[22] Our own research, reported earlier in Chapter 4, provides some evidence that, over recent U.S. history, earnings of scientists have grown at a slower rate than those of other high-status occupational groups, such as medical doctors and lawyers. In the following analysis, we wish to test whether, over three cohorts of students, talented youth became less likely to pursue scientific fields of study. Although this analysis is similar to a parallel analysis reported in Chapter 6, the attainment of a science degree studied in this chapter is a far more concrete indication of pursuit of a science career than the intention of majoring in science, which was covered in Chapter 6.

Cohort Trends in Attaining Bachelor's Degrees in Science: A Decomposition Analysis

Corresponding to the two-step conceptualization discussed earlier, we focus on two key educational outcomes: attainment of a bachelor's degree within eight years of graduation from high school and attainment of a science degree given a bachelor's degree.[23] We compare three cohorts of high school seniors: 1972, 1982, and 1992. Having already introduced the data on these three cohorts in Chapter 6, we now analyze their actual, rather than expected, degree attainment.[24]

These analyses allow us to add to the results already presented in this chapter in two ways. First, trends in scientific degree production are the result of both trends in degree attainment, regardless of field, and trends in the choice of scientific fields of study among those who attain a given degree. Our analyses allow us to distinguish trends in these two processes. Furthermore, we are able to track trends in the educational attainment of particular subgroups of interest, such as underrepresented minorities.

We first examine overall trends in the two outcome variables, presented in Table 7.1. The top two rows give the unadjusted trends across cohorts in the likelihood of receiving a bachelor's degree. The fraction of men receiving a

Table 7.1. Bachelor's degree and science major attainment, by gender and cohort (percent)

	Male			Female		
	1972 cohort (NLS-72)	1982 cohort (HS&B)	1992 cohort (NELS)	1972 cohort (NLS-72)	1982 cohort (HS&B)	1992 cohort (NELS)
Bachelor's degree	27.8	30.7	30.5	23.9	29.8	36.9
Among top 25% in math achievement	54.5	61.2	64.3	53.5	70.4	75.9
Science major given bachelor's degree	28.7	31.4	28.3	10.2	13.7	13.2
Among top 25% in math achievement	36.9	41.5	38.8	15.7	20.9	19.3
Science subfields given bachelor's degree						
Physical science	7.4	3.4	3.1	3.6	1.3	1.6
Life science	9.6	5.0	8.1	4.6	5.3	8.3
Mathematical science	1.3	2.1	1.6	1.6	1.1	0.9
Computer science	1.1	5.6	3.2	0.1	4.1	0.8
Engineering	9.4	15.6	12.4	0.3	2.1	1.7

Sources: NLS 1972; HS&B 1980; NELS 1988 (see Appendix B).
Note: The sum of percentages across the science subfields does not exactly equal the percentage of science majors due to double-majoring and rounding.

bachelor's degree rose modestly across cohorts, from 27.8 percent in the 1972 cohort to 30.5 percent in the 1992 cohort. For women, the rise was more substantial, from 23.9 percent in the earliest cohort to 36.9 percent in the latest cohort. Students with high mathematical aptitudes, defined as having scored in the top 25 percent on the mathematics test given in each survey, generally had high rates of completing a bachelor's degree (above 50 percent in all cases). For men in this category, completion rates increased moderately, from 54.5 percent in 1972 to 64.3 percent in 1992. Female students with high mathematical aptitudes increased their rate of degree completion by a larger margin, from 53.5 percent for the 1972 cohort to 75.9 percent for the 1992 cohort.

The next two rows present trends in the likelihood of receiving an S/E degree conditional on having received a bachelor's degree in some field. For men, there is no clear trend, with the fraction of college graduates receiving a degree in science varying between 28.3 percent and 31.4 percent. For women, there is an increase in the pursuit of science across cohorts, rising from 10.2 percent of college graduates in the 1972 cohort to about 13–14 percent in the later cohorts. Thus, for both men and women, these trends contradict the downward trends in expectations for science degrees discussed in Chapter 6. While women have made slight inroads into scientific training over this period, the data show that male college graduates in the most recent cohorts were still more than twice as likely as their female counterparts to receive degrees in S/E fields.

There is little evidence that science suffers a "leaky pipeline" during the college years that disproportionately steers students away from scientific fields and toward nonscientific studies. For the 1992 cohort, the share of S/E majors among bachelor's degree recipients is slightly higher than the share of expected S/E degrees among youth expecting a bachelor's degree. For men, 28.3 percent of college degree recipients received degrees in science, whereas 27.5 percent of high school students expecting college degrees expected to receive them in science. For women, the analogous numbers are 13.2 percent and 10.5 percent.

Again, the healthy trends in the science share of degrees received, reported in Table 7.1, seem to contradict a finding in Chapter 6 showing a declining trend in the science share among degrees expected. We note that teenagers' expectations of their future educational outcomes are full of noise, as evidenced by the fact that a large portion (13–33 percent) of students in our data expected but failed to receive a bachelor's degree. We also know that many students shift into and out of science, especially around the time of entering college.[25] Some students may also change their majors to science as they prepare for nonscientific professions, such as clinical medicine. In any event, the data in Table 7.1 show the continuing strength of scientific training over the cohorts.

The pattern of steady participation in science by male college graduates and increased participation by female college graduates also holds true among students with high mathematical aptitudes. High-achieving men's likelihood of receiving S/E degrees was 36.9 percent in 1972 and 38.8 percent in 1992; high-achieving women's likelihood increased from 15.7 percent for the 1972 cohort to 19.3 percent for the 1992 cohort. These results, too, diverge from those concerning the students' expectations presented in Chapter 6, which suggested a moderate decline. In particular, we find no evidence that college graduates with high aptitudes for math have gravitated away from scientific fields.[26]

We further disaggregate the trend data on science degrees by field and present them in the last panel of data in Table 7.1. Consistent with results reported in Figure 7.1, for both men and women there is some evidence of declining pursuit of physical science degrees. In the 1972 cohort, 7.4 percent of male college graduates and 3.6 percent of female college graduates received a degree in physical science, but the comparable percentages in the 1992 cohort were only 3.1 percent and 1.6 percent. Offsetting this decline in the pursuit of physical science was an increase between the 1972 cohort and the later cohorts in the fraction of students receiving degrees in engineering. For men, the fraction of college graduates receiving a bachelor's degree in engineering rose from 9.4 percent in the 1972 cohort to 15.6 percent and 12.4 percent in the later two cohorts, respectively. For women, the analogous numbers rose from 0.3 percent to 2.1 percent and 1.7 percent. While engineering is the largest subfield of science majors for men and accounts for more than 10 percent of all male college graduates in the later two cohorts, it remains a far less common pursuit for women, never capturing more than 2.1 percent of a graduating cohort. Consistent with the results for students' expectations shown in Chapter 6, the 1982 cohort was a high point for engineering.

Women's gains in attaining S/E degrees over the cohorts were concentrated in life science, the most popular scientific field for women. The fraction of female college graduates receiving a degree in life science rose from 4.6 percent in the 1972 cohort to 8.3 percent in the 1992 cohort. Pursuit of math degrees has always been less common, with the share of degrees in mathematical science never much above 2 percent for either sex. Across the cohorts, the share stayed fairly constant for men and declined for women.

Analogously to Chapter 6, we explore racial differences in degree attainment in the 1992 cohort.[27] Without any statistical adjustments, the raw comparisons indicate large racial differences: Asians are most likely, and Hispanics and African Americans least likely, to have received degrees in science, with whites in the middle. We are again interested in the extent to which these racial differences are attributable to racial differences in bachelor's degree attainment, as opposed to major choice among those receiving a bachelor's degree. We use the same set of demographic and achievement predictors that we used in Chapter 6 to model these two outcomes.[28] Furthermore, we test whether the differences are statistically significant once other differences are controlled for. As we did in Chapter 6, we focus on the results for the most recent cohort, although the results for the earlier cohorts are similar. For both men and women, in the demographic models that exclude measures of achievement, there are significant racial differences in the likelihood of receiving a bachelor's degree: Asians and whites are more likely than Hispanics and African Americans to receive bachelor's degrees. Once achievement scores are controlled for in the score model, no significant racial differences remain for either sex. Among college graduates, there are significant racial differences in the likelihood of having pursued a scientific major in both the demographic and the score models. With or without controlling for achievement scores, Asians of both sexes are the most likely to major in scientific fields. Among women, African Americans are also significantly more likely than white women to major in a scientific field.

As before, we also examine differences by the education of students' mothers and the incomes of their families while the students were in high school. The results are similar to those presented in Chapter 6. Maternal education and family income are positively associated not only with expectations of receiving a bachelor's degree but also with the likelihood of doing so. As expected, the inclusion of controls for achievement reduces but does not eliminate the association, indicating that some but not all of the relationship between privileged family origins and educational attainment operates through its positive association with achievement, which in turn is associated with college degree completion. There is no significant association between either family income or maternal education and receipt of a science degree among college graduates, for either men or women, with or without the inclusion of the achievement controls. That is to say, family resources appear to operate primarily to give children access to higher education, rather than to influence their choices regarding science fields of study.

Hence, consistent with our two-step conceptualization, causal mechanisms are different for the attainment of a bachelor's degree than for the choice of a science degree given a bachelor's degree. While the attainment of a bachelor's degree is highly dependent on a family's socioeconomic resources, the choice of a science degree is not influenced much by family background factors, with both outcomes strongly affected by academic performance. Thus, it would be incorrect to infer that science degrees are an elite class of degrees open only to children from privileged backgrounds. Rather, entry to science is open to all who perform well in science, provided that they have access to higher education. Of course, access to higher education and high performance are no small conditions, as social origin is a significant determinant of both. However, this result indicates that the lower representation of students from disadvantaged backgrounds among the scientific elite is likely due to more general stratification in access to education, rather than being distinctive to scientific training. This result is consistent with the universalism hypothesis, suggesting that science is open to children from all family backgrounds, conditional on achievement and access to education.

Our results also demonstrate the crucial role of academic performance in determining both the attainment of a bachelor's degree and the choice of a science degree given a bachelor's degree. We observed that measures of academic achievement in math influence both the attainment of a bachelor's degree and the choice of a degree in science among those who complete bachelor's degrees. This finding suggests that persons with good math backgrounds are able to capitalize on their math ability in pursuing scientific education in college. However, we found much less evidence that students' math scores relative to their reading scores predict the receipt of scientific degrees. This is consistent with the results in the previous chapter that absolute math achievement, but not math achievement relative to reading achievement, was predictive of expecting a science degree.

Why Do Students Leave Science?

The analyses presented here and in the previous chapter tell us something about which students plan to study science and which actually complete bachelor's degrees in science. Furthermore, the results reveal that far fewer students actually complete scientific training at the undergraduate level than expected to do so, the primary reason being that fewer students complete bachelor's degrees than expected to, not that science is a smaller share of realized degrees than expected degrees. However, these results tell us little about why students who were once interested in scientific education fail to complete science degrees. To explore this question, we turn to the results of a three-year ethnographic study of students from seven different institutions of higher learning in the early 1990s. In their 1997 book, *Talking about Leaving,* Seymour and Hewitt explored a wide range of factors that may be implicated in students' decisions to leave science, math, and engineering (SME) majors. The interviews in the study yielded a list of twenty-three categories into which the issues raised by participants fell. Participants were labeled as either "switchers" (those who left science majors) or "non-switchers" (those who remained in them).[29]

Is switching out of science caused by challenging undergraduate science coursework that serves the function of gatekeeping? The findings are not so simple. Although 34 percent of switchers had "felt discouraged/lost confidence due to low grades in early years," as opposed to 12 percent of nonswitchers, the two groups differed by a mere 2 percent in having experienced "conceptual difficulties with one or more SME subjects" (switchers = 27 percent and non-switchers = 25 percent). Seymour and Hewitt attributed much of the difference regarding discouragement to differences in students' "attitudes or coping strategies" in response to low grades rather than to grades actually being lower. Overall, they concluded that academic performance differences played a relatively minor role in switching versus non-switching.[30] The concerns that were most commonly cited by both switchers and nonswitchers mainly had to do with the quality of their educational experiences, both past and present. Students complained about poor teaching, harsh "weeding-out" grading policies, excessive workloads, inadequate preparation during high school years, and inadequate counseling and support.

Thus, it is not clear that switching out occurred disproportionately among students with lower levels of academic achievement in science. Instead, students' noncognitive skills—their "coping strategies"—appear to have played a significant role in determining their persistence in science.[31] Those students who successfully receive scientific training are thus selected at least in part on the basis of noncognitive traits. To some extent, this selection mechanism is useful, as students who persevere despite academic challenges are more likely to pursue scientific work with persistence, especially when the rewards are not immediate. Similarly, students who are able to recover from a poor performance on an exam are likely to be able to cope with and respond to negative feedback later in their science careers. Unfortunately, this selection system may result in losing high-achieving students to other fields who might succeed in science had they persisted in spite of temporary setbacks in college.

Switchers also experienced "lack of/loss of interest in SME: 'turned off by science'" and feelings that "non-SME major offers better education / more interest." These factors were the most common reasons given by switchers for their decisions (43 percent and 40 percent, respectively), and "lack of/loss of interest" could be attributed to uninspiring teachers, making this an issue also relating to educational quality. In addition, 82 percent of switchers and 40 percent of non-switchers felt that their "reasons for choice of SME major" had proved "inappropriate."[32] Some of these concerns echo those described in Chapter 6 in regards to youth who simply do not find science interesting. For the students in Seymour and Hewitt's study, this lack of interest surfaced at a later point—all of the subjects at least at one point intended to major in a scientific field. The fact that so many of those leaving science believed that their original reason for wanting to major in science was inappropriate is telling. If the goal of a government policy is to increase the number of young people with advanced training in science, it is not enough simply to increase interest in science among youth; this interest must have a foundation that will sustain the student through the sometimes discouraging period of scientific training.

Transitions to Graduate Programs

A study of the trends in attaining a bachelor's degree has limited implications for an assessment of American science, for its research and development hinge hugely on scientists with advanced degrees. Hence, it is necessary also to understand trends in the transition to graduate programs after completing a bachelor's degree in S/E. Toward this end, we constructed a cross-cohort trend analysis of the post-baccalaureate paths of scientific degree recipients by using two series of repeated cross-sectional surveys—the New Entrants Survey (NES, 1976–1988) and the National Survey of Recent College Graduates (NSRCG, 2003–2006). Both series of surveys were conducted by the National Science Foundation, and our analysis of these data is described in more detail in Appendix C. The target population of the surveys consists of individuals who received a bachelor's or master's degree in the United States within two years (NES and NSRCG 2003) or three years (NSRCG 2006) prior to the survey year. Taken together, the data provide a three-decade trend (1976–2006) of transitions from bachelor's degrees in science to advanced degrees and occupations.

To properly interpret transition data, it is useful to refer to a simple flow chart, Figure 7.3, which organizes six potential destinations (called states)

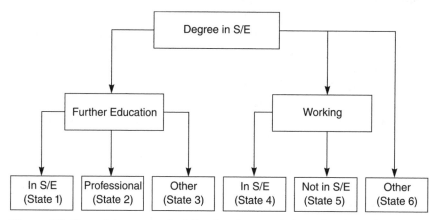

Figure 7.3. Potential career paths following the completion of a science/engineering (S/E) degree.

into hierarchical order with a set of four questions.[33] We first consider, after a person graduates with a degree in S/E, whether the person is either in graduate school for further education or working in the labor force. If the person is not engaged in either of these activities, he or she belongs to state 6. For those not in state 6, we ask whether a person is pursuing further education or working in the labor force. For someone pursuing further education, we next ask whether the student is in an S/E program (state 1), in a professional program such as law and medicine (state 2), or in a non-S/E graduate program such as social science (state 3).[34] For a person working in the labor force, we ask whether the occupation of employment is in S/E (state 4) or non-S/E (state 5). We use Figure 7.3 for our analysis of the NES and NSRCG data reported in this chapter and in Chapter 8. We measure transition states consistently, two or three years after the completion of a degree in S/E, at either the bachelor's or the master's level.

Table 7.2 shows trends over time in transition rates to graduate programs for men and women in the NES and NSRCG data.[35] Note a severe data limitation in the NES data, which did not collect field of study for persons who were in graduate programs. Therefore, for the years from 1976 to 1988, we are not able to separate outflows to graduate programs in S/E (state 1) from outflows to professional programs or graduate programs in other fields (states 2 and 3). Entries for these years represent transition rates to all graduate programs regardless of field (i.e., state 1 + state 2 + state 3).[36] For the NSRCG data, we are able to separate S/E graduate programs from non-S/E graduate programs and professional programs and will present these state-specific transition rates later in Table 7.3.

Let us look first at the broad trends in overall transitions to graduate programs among recipients of a bachelor's degree in S/E, shown in the first panel. We observe an overall decline in continuation rates in the earlier years in the series, for both men and women. While about 44.8 percent of both men and women receiving a bachelor's degree in science between 1974 and 1975 were enrolled in graduate study in 1976, these rates declined to around 30 percent for both in 1988 within the NES series. Due to the possibility that the NSRCG data may not be comparable to the NES data, we exercise caution in comparing trends across the data sets. Still, the continuation rates in the 2003 and 2006 NSRCG data are within the expected range in comparison to the NES series, except that we observe a large gender difference in the NSRCG data, with female recipients of bachelor's degrees in science far more likely to continue to graduate education than their male counterparts (35.4 percent versus 29.8 percent in 2003 and 35.3 percent versus 26.0 percent in 2006).

At the master's degree level, we also observe a generally declining trend. For men, the continuation rate from a scientific master's degree to further

Table 7.2. Transition rates from science degree to further education, by gender and cohort (percent)

Level of degree attained in past two years	NES							NSRCG	
	1976	1980	1982	1984	1986	1988	2003	2006
Bachelor's degree								
Male	44.8	36.9	33.7	35.5	29.6	30.0	32.0	29.6
	44.8	38.5	32.9	36.1	30.6	29.9	29.8	26.0
Female	44.8	32.3	35.6	33.9	27.3	30.3	35.4	35.3
Master's degree								
Male	31.4	28.5	31.4	29.8	27.1	26.6	28.8	27.3
	31.4	29.0	31.9	30.9	28.0	28.8	30.0	29.0
Female	*31.6*	26.6	29.7	26.0	23.9	19.4	26.2	23.8

Sources: NES 1976–1988; NSRCG 2003, 2006 (see Appendix C).
Notes: Entries are transition rates for inflow into (State 1 + State 2 + State 3). The states are defined in Figure 7.3. Estimates based on fewer than 100 cases are presented in italics.

graduate education fell, from 31.4 percent among 1974–1975 master's degree recipients to 28.8 percent among 1986–1987 master's degree recipients in the NES series. For women, the decline was larger, from 31.6 percent to 19.4 percent. At the master's level, a gender disparity emerged beginning in 1980, men's continuation rate being higher than women's, in the direction opposite that of the gender gap at the bachelor's level (e.g., 29.0 percent for men versus 23.8 percent for women in 2006).

When recent recipients of science degrees at the bachelor's and master's levels continue their education, do they study science? Unfortunately, the NES data do not contain appropriate information pertaining to this question. However, the 2003 and 2006 NSRCG data allow us to separate out recent graduates at both the bachelor's and master's levels who were enrolled in S/E programs (state 1), professional programs (state 2), or other programs (state 3). Although the short interval in the NSRCG series between 2003 and 2006 tells us little about long-term trends, the data are informative as to where recent science graduates go for their further education. In Table 7.3, we present the destination-specific continuation rates by level of degree, year, and gender or field of study.[37]

A few major findings emerge from the data. First of all, we found that among those graduates who pursued further education, a large proportion was enrolled in S/E programs. This pattern is particularly strong for graduates with a master's degree in S/E. For example, out of 28.8 percent of recent recipients of a master's degree in S/E who were enrolled in graduate programs in 2003, a majority (20.6 percent) of this group were in S/E programs. Even among bachelor's degree recipients, close to half (15.5 percent) of the 32.0 percent continuing graduates were in S/E programs. In 2006, the continuation rates to S/E were a little higher at the master's level and a little lower at the bachelor's level.

Second, we also observe large gender differences that put women at a disadvantage in terms of retention in science.[38] At the master's level, the large gender difference in favor of men in further education, observed earlier in Table 7.2, is entirely concentrated in continuation rates in S/E education (22.2 percent versus 17.5 percent in 2003 and 23.6 percent versus 19.0 percent in 2006). Men and women with master's degrees had similar transition rates to professional programs or other, non-S/E, programs. However, at the bachelor's level, we find that women have much higher transition rates to professional programs and other programs. In fact, women's higher rates to these non-S/E destinations account for their overall higher continuation rates than men's beyond the bachelor's level, as women's transition rate to S/E graduate programs was either slightly lower than men's (in 2003) or similar to men's (in 2006). From these results, we derive a plausible explanation for our earlier observation that, in recent years, female graduates with bachelor's degrees in

Table 7.3. Transition rates from science degree to specific education programs, by gender and by field, 2003–2006

Level of degree attained in past two years	2003				2006			
	All (states 1–3)	S/E (state 1)	Professional (state 2)	Others (state 3)	All (states 1–3)	S/E (state 1)	Professional (state 2)	Others (state 3)
Bachelor's degree	32.0	15.5	8.7	7.8	29.6	13.4	9.4	6.7
Gender								
Male	29.8	16.6	7.3	5.9	26.0	13.4	7.6	4.9
Female	35.4	14.0	10.8	10.6	35.3	13.5	12.2	9.6
Field								
Biological	43.7	11.5	19.4	12.8	43.9	12.2	22.4	9.3
Engineering	24.8	18.9	1.5	4.4	23.8	17.5	2.8	3.5
Physical	46.5	29.9	10.7	5.9	42.8	27.0	8.1	7.7
Mathematical	18.8	13.0	0.7	5.1	15.5	7.8	1.3	6.4
Master's degree	28.8	20.6	2.9	5.2	27.3	22.1	1.6	3.6
Gender								
Male	30.0	22.2	2.8	5.1	29.0	23.6	1.5	3.9
Female	26.2	17.5	3.2	5.5	23.8	19.0	2.0	2.8
Field								
Biological	37.8	19.7	12.1	6.1	28.6	19.1	6.2	3.2
Engineering	26.5	21.6	1.3	3.6	27.1	22.7	1.2	3.1
Physical	39.1	34.1	1.2	3.8	47.5	44.2	1.1	2.1
Mathematical	24.1	16.2	0.5	7.3	22.5	17.7	0.1	4.7

Sources: NES 1976–1988; NSRCG 2003, 2006 (see Appendix C).

Note: Entries are transition rates for inflow into State 1, State 2, or State 3. The states are defined in Figure 7.3.

science have been more likely, but female graduates with master's degrees in science less likely, to continue their graduate education than their male counterparts. Although women have increased their commitment to professional paid work at a level surpassing men's, their commitment to science careers, especially at the doctoral level, has not matched the same level.

Third, there is a great amount of variation by field. The continuation rates to S/E programs were highest in physical science, 34–44 percent at the master's level and 27–30 percent at the bachelor's level. Next is engineering, with 22–23 percent at the master's level and 18–19 percent at the bachelor's level continuing S/E education. The lowest continuation rates in S/E were found in biological science and mathematical science, ranging from 16 percent to 20 percent at the master's level and from 8 percent to 13 percent at the bachelor's level. Recipients of bachelor's degrees in biological science were most likely to go on to professional schools. While only 8.7 percent and 9.4 percent (respectively for 2003 and 2006) of recent recipients of S/E bachelor's degrees were enrolled in professional schools, the percentages were 19.4 percent and 22.4 percent (respectively for 2003 and 2006) among recipients of degrees in biological science. In contrast, recipients of bachelor's degrees in engineering and mathematical science had very low rates of attending professional schools (all under 3 percent).

Finally, more detailed results reported in Appendix D revealed that the gender differences shown in Table 7.3 are, in large part, due to sex segregation by major, as women are far more concentrated in biological science than are men.[39] That is, women's higher concentration in biological science partly explains their higher likelihood of transitioning to professional schools. Both men and women in biological science are very likely to transition to professional schools, but men's rates of attending professional schools were significantly *higher* than women's (24–28 percent versus 17–19 percent at the bachelor's level). Among students receiving bachelor's degrees in biological science, women are no less likely to transition to S/E graduate schools than are men.

Our finding from this analysis challenges a popular view that persons with a bachelor's degree in S/E are likely to shift to a non-S/E field if they continue graduate education.[40] Our results indicate that pursuit of a professional degree is less common than is graduate work in S/E: overall, only about 9 percent of recent college graduates with S/E degrees were enrolled in professional schools, compared with 13–16 percent enrolled in S/E graduate school. For master's degree recipients, the disparity is much more pronounced: about 2–3 percent of recent master's degree recipients were enrolled in professional school, compared with 21–22 percent enrolled in S/E graduate programs. However, we do note one exception. Among recipients of bachelor's degrees in biological science, the transition to professional schools (mainly to medical schools) was more likely than the transition to S/E graduate programs, consistent with the stereotype that many undergraduates may complete a bachelor's degree in biology with an explicit career goal of attending a medical school and becoming a doctor.

Clearly, professional schools draw a sizable minority of S/E degree recipients at the bachelor's level away from further study in S/E fields. Overall, however, S/E graduate study remains the dominant destination at the graduate level for S/E bachelor's degree recipients, outcompeting either professional school or other non-S/E graduate study.

We believe that this discrepancy between our results and the popular belief that most graduates of S/E degrees leave science is caused primarily by the loose definition of "science and engineering" used in *Science and Engineering Indicators*.[41] In particular, as shown in the original 2006 NSF report,[42] graduates with bachelor's degrees in social science are unlikely (indeed, least likely across all major fields) to continue graduate education in science (including social science). In addition, U.S. universities now produce a very large number of undergraduate degrees in social science majors. In the year 2006, the number of bachelor's degrees awarded in social science was 236,665, comparable in size to our estimate of 237,350 for the total number of S/E (excluding social science) bachelor's degrees awarded that year. Thus, inclusion of social science makes the overall education continuation rate appear alarmingly low. A careful reading of the original NSF report reveals that it actually stated that graduates with bachelor's degrees in mathematical science, physical science, and engineering were all more likely to continue graduate education in S/E than in non-S/E fields.[43] Therefore, our estimates provide a good description of the rates at which S/E undergraduates transition to S/E graduate programs. When a narrower definition of science is used, there is little evidence that the transition from undergraduate to graduate studies is a period of significant attrition from science to nonscience fields.

Summary

The broad picture that emerges from this chapter is one of relative stability in science education over the last four decades in the United States. None of our results should cause major concern to science policymakers. At least in terms of the production of young adults with college-level scientific training, there is no evidence that America has experienced a sharp decline in recent decades. Still, we have observed some changes that are worth summarizing:

(1) The production of science degrees at all levels has grown, albeit at slow rates.

(2) Women have increased their participation in science education at all levels. The increase is most pronounced in biological science. While women have either narrowed their gap with men or surpassed men by some indicators of undergraduate education and professional training, they still lag behind men at advanced levels of training in science.

(3) A larger and larger share of S/E degrees at the master's and doctoral levels has been awarded to foreign students.

(4) American youths, particularly those with high mathematical aptitudes, continue to complete science education at the bachelor's degree level. The pursuit of science education among those with college degrees is consistent with universalism. With the exception of the overrepresentation of Asians and African American women in S/E fields of study, we see few differences by social origin in college graduates' choices of S/E fields.

(5) The rate at which graduates with bachelor's degrees in S/E continue graduate education has declined from around 40 percent in the late 1970s to around 30 percent since the late 1980s.

(6) However, most graduates of science programs at both the bachelor's level and the master's level still study S/E if they do continue their education, with the exception of recipients of bachelor's degrees in biological science, for whom the likelihood of attending professional schools is higher. Other studies have underestimated persistence in science education when they included social science as part of science.

8

Finding Work in Science

It's all about jobs, stupid.

—AMANDA VANSTONE, AUSTRALIAN POLITICIAN, ADDRESSING HER
COUNTRY'S SENATE, 2006

One reason some scholars and observers question the common assertion about a shortage of scientists is their concern with job prospects for newly trained scientists. According to these scholars and observers, not only is there no shortage of scientists in the United States, there is actually a surplus of scientists, largely resulting from immigration. In a 2010 article provocatively titled "The Real Science Gap," science journalist Beryl Lieff Benderly quoted economist Richard Freeman and other scholars denying that a science shortage exists. Benderly characterized the loud complaints about scientist shortages as "profound irony" at a time when "scores of thousands of young Ph.D.s labor in the nation's university labs as low-paid, temporary workers, ostensibly training for permanent faculty positions that will never exist."[1] Instead of a shortage, opponents such as Benderly and Freeman argue that there is currently an oversupply, or indeed a glut, of scientists.

According to the glut proponents, claims of a scientist shortage encourage practices likely to produce even more young scientists and attract an even greater flow of immigrant scientists at a time when there is already an oversupply relative to available jobs in science. In their view, the deterioration of labor market outcomes (in terms of training-related employment, job security, earnings, and autonomy) among scientists is the real deterrent to talented youths who may aspire to become scientists.

Can we have too many scientists in America? The glut proponents distinguish the interests of today's science stakeholders from those of American science itself in the long term. An oversupply of young scientists may be good for employers, such as universities, high-tech firms, and senior scientists, who can hire junior scientists cheaply to do labor-intensive scientific work today, but poor economic outcomes for young scientists send a negative signal to talented young Americans who are in a good position to choose among competing careers.

Hence, as the argument goes, the lack of economically good jobs is the real threat to the long-term well-being of American science. Put another way, the

115

short-term interest in lowering the labor costs of science today hurts the long-term interest of attracting scientifically gifted youths to science in the future, thus contributing to an eventual shortage of scientists.

What is the evidence for this glut theory? The above-mentioned Benderly article cited three specific points. First, wages of scientists have declined. Second, citing *Science and Engineering Indicators 2008*, "three times as many Americans earn degrees in science and engineering each year as can find work in those fields."[2] Third, young scientists today need to spend many years in temporary and insecure postdoctoral positions before a very few of them become permanently employed researchers.

In Chapter 4, we presented evidence that scientists' earnings relative to those of other highly trained professionals such as medical doctors and lawyers have declined since 1960. Even in absolute terms, scientists' earnings have stagnated. In this chapter, we will review empirical evidence on Benderly's last two points about scientific employment and postdocs. We turn first to an examination of the claim that only one out of three recipients of science and engineering (S/E) degrees can find work in science or engineering.

Only One out of Three?

"The S&E employment of S&E graduates is also a fairly consistent one-third of S&E graduates," claimed a 2007 report released by the Urban Institute in response to the 2007 NAS report *Rising above the Gathering Storm*.[3] A government-sponsored official publication series, *Science and Engineering Indicators*, which is commonly considered the most authoritative source on scientific workforce statistics, has also released estimates that seem to bolster this claim. The 2010 edition, for example, reported a large discrepancy (roughly in the ratio of one to three) between the size of the science and engineering workforce and the number of individuals with a bachelor's degree or higher in science or engineering, with "between 4.3 million and 5.8 million" for the workforce versus "16.6 million" for the individuals with science or engineering degrees.[4]

We analyzed the original data but arrived at a different conclusion. We began with an estimation of the size of the science and engineering labor force. As was the case in Chapter 4, we are concerned only with employed workers. From Table 4.1, we already know that 3.3 percent of the employed labor force in 2006–2008 worked as scientists, excluding social science but including computer science. Given the size of the total employed labor force in 2007 of about 146.0 million,[5] this estimate puts the size of the scientific labor force in 2007 at around 4.8 million. This estimate is consistent with the 4.3–5.8 million range for 2006 reported in *Science and Engineering Indicators 2010*.

We then carried out an estimation of the number of persons in the 2008 American labor force with at least a science or engineering bachelor's degree. Through two different methods of estimation, we found this number to be somewhere between 7.9 and 10.6 million, a number far short of the "16.6 million" reported in the *Science and Engineering Indicators 2010*.[6] In the first method, we simply summed the total number of bachelor's degrees in science and engineering awarded in the United States between 1966 and 2008, reported earlier in Figure 7.1a. This number came to 7.6 million. If a person typically finishes a bachelor's degree at age twenty-two, he or she may not exit from the labor force until age sixty-five, or forty-three years later. Thus, the total number of science or engineering degrees awarded in the past forty-three years is not a bad place to start.[7] To refine our estimates of persons with science or engineering degrees, we took several additional factors into account: (1) immigrants who came to the United States with bachelor's degrees in science and engineering from their home countries; (2) individuals who obtained science degrees at advanced levels after first completing bachelor's degrees in nonscience; and (3) differential work durations in the labor force.

Factor (1), the immigration factor, should increase the estimate significantly. It can be further decomposed into two groups. The first group consists of foreign students who completed bachelor's degrees in their home countries, came to the United States for advanced degrees, and later stayed in the United States. The other group consists of immigrants who completed their educations abroad. We estimated in Table 4.1 that about 27.5 percent, or about 1.3 million, of the scientific labor force was foreign-born in 2007. Of these, about half had attained a college degree abroad.[8] Allowing for immigrants with science degrees who are not working as scientists, we estimate the total size as 0.6 to 1.2 million.[9]

We do not have good data bearing on factor (2), transitions from nonscience to science. However, we do not think that this affects the overall estimate much, because very few individuals are able to make the transition from a nonscience undergraduate background to a science graduate program. Again, we are excluding social science. For convenience, we allow for 0.4 to 0.8 million, or 5–10 percent of degree recipients of science or engineering bachelor's degrees, to make such transitions.

Factor (3), differential work durations in the labor force, can modify our estimate both positively and negatively. Some individuals may work longer than forty-three years after completing a bachelor's degree, increasing the estimate of the target population of potential scientists. However, for many other reasons, such as mortality before retirement, voluntary labor force withdrawal (particularly among women), early retirement, and potential scientists completing a bachelor's degree at ages older than twenty-two, this factor may result in a reduction of the estimate. In our exercise, we allow for up to a 0–5 percent reduction, or from no change up to 0.4 million, due to factor (3).

Other minor factors, such as emigration, may exist, but they should not affect our estimate in significant ways. After considering the three major factors above, we put the estimate of potential scientists with at least a bachelor's degree in science or engineering in 2008 at 8.2 to 9.6 million.

In the second method, we first estimated the total number of workers in the labor force with at least a college degree. From Table 4.1, we know that 32.9 percent of the employed held at least a bachelor's degree in 2007. Extrapolating the growth in the college-educated labor force between 2000 and 2007 beyond 2007, we found that this percentage should have increased to 33.4 percent in 2008. In 2008, total employment was about 145.4 million.[10] Multiplying total employment by the estimated percentage holding at least a bachelor's degree in 2008, we obtained the estimated number of workers with a bachelor's degree at 48.5 million. From degree production data, we know that the percentage of bachelor's degrees awarded in science and engineering fields varied between 15 percent and 21 percent, with a general declining trend.[11] The percentage was only between 15 percent and 17 percent in the years 1989 to 2008. If we used 16–19 percent as a guideline, we would obtain between 7.8 million and 9.2 million workers with science or engineering degrees. Again, several other factors should be considered in modifying this baseline estimate: (1) the share of science and engineering degrees is higher for immigrants with foreign bachelor's degrees coming to the United States than among all individuals receiving bachelor's degrees in the United States; (2) persons may have changed from nonscience fields at the undergraduate level to science fields at advanced levels; and (3) the differential timing of labor force withdrawal. We added 0.1 to 0.2 million additional persons due to factor (1) and kept the same allowances for factors (2) and (3) as in method 1. Last, we considered the fact that some potential scientists may be unemployed. Over the period 1983–2006, unemployment for scientists ranged between 1.3 percent and 4.0 percent.[12] Thus, we estimated the total number of potential scientists in the 2008 labor force to be between 8.0 million and 10.6 million.

The two methods yielded a consistent result, although the second method gives a larger range. It appears that there were about 8.0 to 10.6 million workers with at least a bachelor's degree in science or engineering in the 2008 labor force. We may consider them potential scientists. We already know that the number of scientists and engineers in 2008 according to the occupation-based definition was about 4.8 million.[13] Thus, the ratio between the number of actual scientists and the number of potential scientists is between 45 percent and 60 percent, more than the commonly accepted ratio of "one out of three." In other words, the actual ratio is not one in three but about one in two.

Why is our estimated number of persons with science or engineering degrees so different from the 16.6 million reported by *Science and Engineering Indicators 2010*? The answer is very similar to our discussion of graduate school continuation rates in Chapter 7: what fields of study are included as science and engineering is crucial. In our analysis, we tried as closely as possible to include fields of study that match the occupations that we considered the scientific labor force. We included physical science, life science, mathematical science (including computer science), and engineering. *Science and Engineering Indicators*, by contrast, also includes social science. For the purpose of assessing utilization of science education, we believe that the exclusion of social science is necessary, for a number of reasons. First, even if one wishes to consider social science as science, an undergraduate degree in social science is not closely linked to a social science career. As we discussed in Chapter 7, for most undergraduates, social science is more of a liberal arts than a science major. Second, as an undergraduate major option, social science is much bigger than other fields of science and engineering, and it has grown rapidly in recent years.[14] As previously mentioned, the number of bachelor's degrees awarded in social science was about the same as the total number of science and engineering bachelor's degrees awarded. Finally, the outcry about the state of American science has typically been concerned with the importance and well-being of natural science rather than that of social science.

By including social science majors, *Science and Engineering Indicators* reports statistics on a broader set of degrees. In general, more information is preferable to less, so this decision is not unreasonable by itself. In this particular case, however, the combination of the lack of a close linkage between an undergraduate degree in social science and later employment in social science and the huge size of social science relative to natural science and engineering makes observed data on the match (or mismatch) between science education and science employment misleading. These highly aggregated data severely underestimate the extent to which graduates with science degrees find jobs in science occupations or pursue graduate studies in science, as discussed here and in Chapter 7.

Work after Bachelor's and Master's Degrees

The estimation exercises in the preceding analysis are at best indirect, because they are based on aggregate statistics compiled from different sources and thus different individuals. Any differences across data sources in definitions such as field of study or target population could easily introduce biases to the results. To directly address the issue of utilization of science degrees, it is preferable to use data pertaining to the same individuals, that is, on labor force outcomes of recipients of science degrees. This is the analysis we describe in this section.

Because we are interested in trends, we analyzed survey data for multiple years that are roughly comparable over time, beginning in 1976. As we did in Chapter 7, we drew data for this analysis from two different survey series: the New Entrants Survey (NES) for the period 1976–1988 and the National Survey of Recent College Graduates (NSRCG) for the period 2003–2006.[15] One caveat is that due to irresolvable discrepancies in the coding schemes for fields of study and occupations across the two series, end results are not strictly comparable.[16] In analyzing these data, we focused on a common metric: the proportion working in science and engineering occupations among those who were employed within two years of receiving a degree in science or engineering at the bachelor's or master's level.[17] We intend this metric to measure the rate of utilization of scientific training among recent science graduates in the labor force. In Table 8.1, we show the results for this metric by degree level, gender, and year (thus data source).[18]

Throughout the period, the utilization of science degrees for employment was much higher at the master's level than at the bachelor's level for both men and women. For example, in 1976, 84 percent of male recent recipients of master's degrees were working in science, compared with 64 percent among those with bachelor's degrees. This difference makes sense because a master's degree in science means more specific scientific training and thus should be tied more closely to science-related employment. We observe a sudden decline

Table 8.1. Transition rates from science degree to science occupation, conditional on working, by gender and cohort (percent)

Level of degree attained in past 2 years	NES						NSRCG	
	1976	1980	1982	1984	1986	1988	2003	2006
Bachelor's degree	60.9	72.7	82.5	81.9	80.4	81.5	54.1	54.0
Male	63.9	76.9	86.0	85.6	83.9	84.5	62.7	64.4
Female	47.1	49.0	71.8	73.2	72.0	74.1	40.0	34.0
Master's degree	82.5	86.9	89.2	90.7	90.1	88.8	80.8	76.9
Male	83.5	83.7	90.1	92.3	91.6	91.3	85.1	81.0
Female	*74.8*	60.3	85.7	85.3	84.9	80.9	71.8	68.0

Sources: NES 1976–1988; NSRCG 2003, 2006 (see Appendix C).

Notes: Entries are transition rates calculated as (State 4)/(State 4 + State 5). The states are defined in Figure 7.3. Estimates based on fewer than 100 cases are presented in italics.

in the utilization rate from the end of the NES data (1988) to the beginning of the NSRCG data (2003), the pattern being more pronounced at the bachelor's degree level than at the master's degree level. At the bachelor's level, the rate dropped from 85 percent to 63 percent for men and from 74 percent to 40 percent for women. At the master's level, the rate dropped from 91 percent to 85 percent for men and from 81 percent to 72 percent for women. Without data on the intermediate (i.e., 1988–2003) period, we cannot be certain how much of the observed decline is attributable to real social change and how much to incomparability between the two data sources. If a decline did occur, which seems plausible given the very large size of the changes, it would support a complaint by glut proponents that it has become increasingly difficult for graduates with science education to find jobs in science.

The NES data covered enough years for us to examine trends between 1976 and 1988 without worrying about compatibility issues across data sets. From the NES data, we first observe that the utilization rate increased from 1976 through 1982 and then became relatively stable in 1982–1988. In 1976, the utilization rate of science degrees for recent bachelor's degree recipients was 64 percent among males and 47 percent among females. By 1982, those figures had risen to 86 percent among males and 72 percent among females. From 1982 to 1988, the utilization rates were stable, ranging between 84 percent and 86 percent for men and between 72 percent and 74 percent for women. Similarly, utilization rates among those with a recent scientific master's degree rose from 84 percent for men and 75 percent for women in 1976 to 90 percent for men and 86 percent for women in 1982, followed by a period of relative stability.

For the NSRCG data, the 2003–2006 period was not long enough for us to observe trends. Again, the observed utilization rates from the NSRCG data in this recent period were much lower than for the earlier period. Among recipients of bachelor's degrees, the rate fluctuated between 63 percent and 64 percent for men and between 34 percent and 40 percent for women. Among master's degree recipients, it was between 81 percent and 85 percent for men and between 68 percent and 72 percent for women.

Overall, the employment data shown in Table 8.1 contradict the assertion that "the S&E employment of S&E graduates is also a fairly consistent one-third of S&E graduates."[19] Even for the most recent data in 2003–2006, more than half of all graduates with science degrees at either the bachelor's or the master's level hold scientific occupations if they work at all.[20] Thus, we conclude from these numbers that more than one in two potential scientists in the labor force are now employed as scientists, although the rate was closer to two out of three in the 1980s. Of course, the utilization rate would be much lower if we had included social science majors in our calculation.[21]

Without further information, it is not possible to determine whether the individuals employed in nonscience jobs voluntarily chose nonscience careers or were forced to take nonscience jobs due to the difficulty of finding science-related jobs. One estimate puts the percentage of involuntary nonscience employment among recipients of science degrees at the bachelor's level at less than 10 percent.[22] A science bachelor's degree, in particular, might well be an intended step on a path to a nonscientific job in, for example, a health profession.

What about unemployment? If considerable numbers of potential scientists are unemployed, then the preceding calculations, which are based on those who are employed, could mask the effects of a glut of scientists. However, the unemployment rates of scientists are consistently lower than those of the labor force in general and of members of the labor force with a bachelor's degree or higher.[23] Again, there is no evidence that an oversupply of scientists is leading to high unemployment for this group.

Doctoral Degrees and the Academic Job Market

A doctoral degree is a prerequisite for most academic jobs in contemporary America. By "academic" we mean a work environment within an institution of higher education. However, not all doctoral recipients in science and engineering end up working in academic jobs. About half of them do, and the share of doctoral scientists and engineers working in academia has declined in recent decades.[24] In 2006, 41 percent of doctoral scientists and engineers were employed in four-year universities. Across fields of study, mathematical scientists other than computer scientists were most heavily concentrated in the academic sector (at 59 percent), and engineers the least concentrated (28 percent), with biological scientists similar to mathematical scientists (at 51 percent) and physical scientists similar to engineers (at 34 percent) in this respect.

Relative to the estimated size of the total science and engineering labor force of 4.8 million (for 2007), the academic sector is a tiny fraction, at 2 percent.[25] However, this tiny fraction is responsible for almost all the research output in the form of scientific publications. Indeed, if we were to examine the productivity data of scientists, we would undoubtedly discover that the bulk of scientific output has been published by a tiny fraction—the most productive fraction—of this tiny fraction of the total scientific labor force.[26] By nature, academic science values and rewards research excellence over other things (such as teaching and service) and thus implicitly encourages inequality in research output as well as reward, with a small fraction of scientists receiving most of the recognition over the rest of the field.

Of course, academic employment has never been the only option for doctoral scientists. The next most important sector of employment for them is for-profit private industry. In 2006, almost an equal percentage, 38.3 percent, of doctoral scientists and engineers were employed in this sector.[27] One major advantage of employment in for-profit private industry is that doctoral scientists there enjoy an earnings premium as large as 50 percent over their counterparts in academia, and engineers earn a premium of about 30 percent.[28] Nonetheless, some individuals may prefer academic science for nonfinancial reasons, including the prestige of academic positions, the opportunity to contribute to scholarly work, and greater autonomy in work scope. The following is from *Academic Scientists at Work*, a book giving advice to young scientists:

Perhaps the greatest benefit of academic life is the freedom to work in any field you choose. The caveat is that you will have to find your own funding for your work, convince others that the work is worthwhile for you to stay in the department (tenure issues), and be willing to receive lots of objective—and maybe even subjective—critiques of your ideas. Freedom has its price.[29]

Has it become increasingly difficult for scientists to find academic jobs? This is the concern expressed in many reports, particularly about biomedical fields, on the current state of science.[30] Analyzing data on doctoral scientists employed at academic institutions, we found that academic positions for scientists and engineers grew steadily from 82,400 in 1973 to 188,900 in 2006, at an annualized rate of 2.5 percent, a small rate but still ahead of the increase rate of S/E doctoral degrees (1.5 percent) for the same period.[31] However, these academic employment data may be misleading in that they contain nontenure-track and postdoctoral ("postdoc") appointments—positions that lack both independence and security and thus are not truly academic jobs in the conventional sense. In examining trends among doctoral scientists employed at universities between 1973 and 2006, we do observe a worsening trend in the academic job market for scientists and engineers in the United States. First, the fraction of full-time faculty declined from 86.0 percent in 1973 to 71.0 percent in 2006. Second, the share of postdocs increased, from 4.7 percent in 1973 to 11.1 percent in 2006. Third, the share of nonfaculty employees also increased sharply from 7.0 percent in 1973 to 13.6 percent in 2006.[32] Much of the recent debate concerning the academic labor market is indeed centered around postdocs, particularly in biomedical fields.

Postdocs: Ladder Climbing or "Disguised Unemployment"?

The National Postdoctoral Association defines a postdoc as "an individual holding a doctoral degree who is engaged in a temporary period of mentored research and/or scholarly training for the purpose of acquiring the professional skills needed to pursue a career path of his or her choosing."[33] Postdoc positions are designed to provide additional training in specialized fields to doctorate recipients, but they may also prolong the educational process and discourage some talented youths from pursuing a science or engineering educational path.

The postdoctoral position traces its roots all the way back to the 1870s, when technical apprenticeships became prominent throughout Europe.[34] In America, the overall growth in postdoc positions has not been planned by federal policies, and it has instead resulted from unexpected economic and political factors.[35] The first boom in the American postdoc population occurred in the 1950s, as the onset of the Cold War greatly stimulated federal spending in the sciences.[36] In the 1970s, weakened PhD markets and a significant increase in the supply of researchers due to an influx of foreign students may have forced many science and engineering graduates into postdoc positions.[37]

An ever-increasing number of young scientists now find themselves in these positions. From 1993 to 2006, the number of postdoc appointments in science (excluding social science) increased from 33,000 to 48,000.[38] Also reflecting the increasing importance of postdocs, the fraction of science and engineering doctorates with definite postgraduation plans who were planning postdoctoral study jumped from 46.8 percent to 52.3 percent between 1988 and 2008.[39] The postdoc system has clearly become a prominent feature of the current scientific landscape in America.

What is new about postdocs in recent years is that a postdoc appointment has now evolved into a virtual prerequisite for entry to many faculty positions.[40] This is particularly true in biochemistry fields, where the postdoc—as opposed to the conventional PhD—has become the main proving ground for young scientists.[41]

It is clear that today's postdoc system is a product of American "Big Science," in which many scientists work in research teams backed by large amounts of government funding. Within Big Science, divisions of labor exist within each research team, usually with a senior researcher (PI, principal investigator) applying for research grants, supervising lab work, and presenting and writing papers. Junior researchers, either as postdocs or "soft-money" research staff, work for relatively low pay and status and are often given credit as coauthors on scientific papers. The attraction, of course, is the expectation that postdocs will learn from such practices and later move on to regular faculty positions and run labs themselves. Thus, a postdoc experience is sometimes seen as one integral part of the career cycle of a successful scientist.

Three major criticisms have been expressed regarding the current situation of postdocs in science.[42] The first criticism is that the system stifles creativity. Many of the greatest breakthroughs in the history of science were made by scientists in their youth (say, before age forty).[43] If today's young scientists are kept as postdocs who work mostly on others' old ideas, they have no opportunity to develop independent and innovative research programs for themselves. One piece of worrisome evidence is that the number and share of research grants awarded to young scientists have significantly declined. According to a study sponsored by the National Research Council, awards to researchers ages thirty-five and younger declined by more than 75 percent between 1980 and 2005, despite more grants being awarded overall."[44]

The second criticism is that postdocs are being exploited for the benefit of their employers, senior researchers, and universities. This criticism is well articulated in the writings of science writer Beryl Lieff Benderly, who characterizes today's postdoc appointments as "disguised unemployment."[45] Freshly trained, ambitious, and energetic postdocs provide the best labor for doing bench work in well-funded labs. They are poorly paid but expected to put in long hours.[46] The NIH issues guidelines for National Research Service Award postdoctoral stipends, and these guidelines are highly influential for the setting of postdoctoral salaries by other institutions as well.[47] In 2010, stipends ranged from $37,740 for a postdoc with no prior postdoctoral experience to $52,068 for a postdoc with seven or more years of experience.[48] Making an analogy between postdocs and migrant agricultural workers, Benderly quoted the director of postdoctoral affairs at one stellar university: "The main difference between postdocs and migrant agricultural laborers . . . is that the Ph.D.s don't pick fruit."[49] Of course, the system would not appear unfair if all postdocs could later find jobs as regular faculty and run their own labs with their own postdocs. However, there simply are not enough faculty jobs to go around. The supply of postdocs far exceeds the demand for research faculty, especially in biomedical fields. Thus, most of today's postdocs are unlikely to become tomorrow's professors of science.

The third criticism of the postdoc system is that it discourages young talents from pursuing science careers, which over the long term is likely to lower the quality of scientific personnel. For youths aspiring to become scientists, this new postdoc requirement means an increase in the length of the typical education trajectory and also greater uncertainty regarding career outcomes. In addition to graduate study and PhD work, many students now expect to complete a postdoc term before obtaining a faculty position. The length of the postdoc term itself has also increased throughout the years. In fields where the postdoc is considered especially vital, such as the biosciences, the length of a postdoc term may reach beyond five years.[50] Due to these lengthy postdoc terms, the demographics of the postdoc population are changing. Many postdocs are now in their mid- to late thirties and do not begin independent research until their early forties. These numbers may well transmit discouraging signals to American college students, who must consider whether they want to be continuing their education at an age when most people are married and raising families. Although the issue has not been thoroughly studied, it has been suggested that the increasing length of study deters top students from entering science careers.[51]

Work after Doctoral Degrees

The current postdoc and funding systems may have the unintended effect of impeding the careers of young scientists. However, a lengthening of the path to a faculty position by itself is no proof for glut proponents' suggestion that there exists a severe underutilization of science and engineering education. A 2007 Urban Institute report states that "there is already a large pool with sufficient human capital who choose not to enter, or subsequently leave after entering an S&E job, representing substantial private and public educational investment that is not being utilized."[52] Earlier in this chapter, we corrected a misconception that only one out of three degree holders in science and engineering at the bachelor's and master's levels actually works in science or engineering fields. Now, we will discuss the issue of utilization of doctoral training.

As we discussed earlier, a clear demarcation of training at the doctoral level from that at the bachelor's or master's level is that doctorate recipients are trained to do research and teaching rather than just perform routine tasks. Thus, for recipients of doctoral degrees in science or engineering, we chose a different indicator to measure the utilization of the true intent of their training: whether their primary or secondary work activities are in research and development (R&D) or teaching. We assume that individuals are utilizing their doctoral training if they either contribute to scientific discovery or teach science to others. This assumption is, of course, only partially correct, as doctoral scientists performing other types of activities—management and administration, for example—may also utilize their training fully. As before, we examine trends in this utilization indicator among recent degree recipients. The data are drawn from selected cohorts between 1976 and 2006 of the SED, an annual census of those earning research doctorates.[53] The analyses are limited to individuals who indicated that they had definite commitments for employment or postdoctoral study, excluding those who had not yet committed to a particular employer or institution.

Table 8.2. Percentage of science doctorate recipients whose primary or secondary work activities are R&D or teaching, by field and cohort, 1976–2006

	1976	1981	1986	1991	1996	2001	2006
Total	81.7	80.7	81.1	76.1	70.9	70.4	72.8
Biological	78.5	77.0	75.5	72.0	70.2	69.0	68.6
Physical & mathematical	86.1	84.3	84.5	78.7	74.2	74.0	79.0
Engineering	80.4	81.0	83.6	77.1	69.0	68.8	71.1

Source: Hoffer et al. (2007).
Note: Analysis is restricted to doctorate recipients with definite plans for postgraduation employment.

We present the results in Table 8.2. Overall, there is some evidence of declining utilization among scientists and engineers who recently completed their doctoral degrees. While 82 percent of the 1976 cohort was engaged in R&D or teaching, this rate was only 73 percent in 2006. This declining trend in utilization is not limited to a particular subfield. Utilization in biological science fell from 79 percent in 1976 to 69 percent in 2006. We observe comparable declines from 86 percent to 79 percent in physical science and mathematical science and from 80 percent to 71 percent in engineering. On the whole, however, a vast majority (between 70 percent and 82 percent) of scientists or engineers who recently earned their doctoral degrees were engaged in research or teaching throughout the period.

Summary

Five main findings are presented in this chapter:

(1) The utilization of science education for work has been severely underestimated mainly due to the inclusion of social science majors in past studies. It is not one out of three, as previously claimed, but between one in two and two out of three holders of science degrees who are employed in science-related occupations.

(2) At the bachelor's level, utilization of science education increased between 1976 and 1982, and this increase was most pronounced among women. Utilization of undergraduate science education was stable between 1982 and 1988.

(3) Utilization of science education at the master's level has always been high throughout the past three decades. There has been little change.

(4) The current postdoc system may hamper the career opportunities of young scientists and discourage talented youths from pursuing science careers.

(5) Utilization of science education at the doctoral level has also been high. However, there has been some decline in the share of new science doctorates performing work tasks related to research and teaching.

An overarching conclusion of this chapter is that the labor market situation for scientists in the United States has remained healthy in recent decades, especially at the bachelor's and master's levels. There may be reasons for concern among young doctoral scientists and engineers who aspire to be academic researchers. This group is extremely important if American science is to retain its prominent role in the world, as academic researchers are responsible for producing most of the scientific knowledge. However, they constitute only a tiny fraction of the entire science and engineering labor force. Thus, an assessment of the labor market conditions of academic scientists should be separated from an assessment of the overall science and engineering labor force in the United States.

Conclusion

We accumulate knowledge rather by subtracting error
than by adding truth.
—OTIS DUDLEY DUNCAN, 1994

This book was motivated by a policy debate about the current and future states of American science. As described in the Introduction, the alarmist view, well represented in the 2007 NAS report *Rising above the Gathering Storm*, asserts that the United States is in desperate need of more and better trained scientists to meet new challenges posed by an increasingly globalized and competitive world.[1] The opposing view, discussed in Chapter 8, argues that, far from facing a shortage, the United States has a glut of young scientists with weak employment prospects. A broader debate is over whether we should be concerned that American science is in decline, as summarized in two other reports critiquing the NAS report.[2] In this book, we have carefully evaluated the claims made previously by various authors with the best data currently available to us. Although our evidence does not support the claims of either a glut of young scientists or an impending shortage, we found causes for concern in some areas of American science.

The Big Picture

Is American science in decline? Our answer is a qualified no, based on a body of empirical evidence regarding its various aspects as presented in the preceding chapters. Taken as a whole, the evidence dispels the pessimistic myth that, relative to its recent past since the 1960s, American science is in serious trouble. Specifically, we have found the following:

(1) The scientist labor force in the United States has grown in size (Chapter 4).
(2) Public interest in and support for science in American society have remained high (Chapter 5).
(3) American high school students are doing more course work and performing better in mathematics and science than in the past, although their interest in attaining science education has shown a moderate decline (Chapter 6).

129

(4) American universities have been producing new graduates in science at the bachelor's, master's, and doctoral levels in increasingly large quantities, although the number of science degrees awarded to native-born men has been stable (Chapter 7).

(5) Most graduates with science degrees in the United States have found jobs related to their training, contrary to the glut proponents' claim that there are already too many scientists in America today (Chapter 8).

At the same time, however, we have also found evidence that should cause concerns about American science and thus must qualify our initial conclusion. Our concerns relate mainly to three aspects of American science: earnings, academic science, and international competition. Although these three aspects may be related,[3] we lack the requisite data to establish causality between them and will thus discuss them separately.

The Earnings Factor

An analysis of earnings, reported in Chapter 4, shows that those of scientists have fared poorly since the 1960s. On an absolute scale, scientists' earnings have virtually stagnated after adjusting for inflation. On a relative scale, scientists' earnings have significantly declined in comparison to those of some other high-status, high-education professions (such as medicine or law). Given an increasing trend in economic returns to education and skill in the general U.S. labor force since the 1980s,[4] this is a surprising and perplexing phenomenon, one that may well signal a less healthy state of American science today than previously, as declining earnings should have a negative effect on the number of individuals who decide to become—or to remain—scientists.

Academic Science in Decline?

Academic scientists make up only a small proportion of practicing scientists in the United States, but their positions as leaders in scientific research make their fate important for the future quality of American science. Thus, we need to be concerned about the fact that the share of American science doctorates going on to academic positions has declined. Furthermore, among those in academia, there has been an increase in postdoctoral appointments and a decline in full-time faculty positions. As we shall see in the next section, newly minted American science doctorates face increased competition from scientists trained abroad. Thus, the postdoctoral period lengthens the time until young scholars become independent researchers, while the likelihood of transitioning to a tenure-track position has declined. The heightened costs of scientific training, combined with heightened competition for rewards, may make the already high-risk enterprise of scientific careers in academia unappealing to many young Americans.

The current practice of American science, particularly academic science, is also not without critics within the scientific community. The main focus of their critiques is the fear that the current funding system at the federal level, notably at the NIH and the NSF, discourages innovation. As we discussed earlier, innovation is a key feature of science, as important scientific advances are made only by going beyond conventional wisdom. Why would the current funding model discourage innovation?

As we explained in Chapter 1, American science became Big Science in the post–World War II period, in which many researchers began teaming up to work on large, well-funded projects. In such a system, junior scholars are typically hired as postdocs or research associates to deliver results on large projects funded by a grant made to a senior researcher. Such projects are typically funded, however, only after being approved by committees of other senior researchers operating within an evaluation system called "peer review." The peer-review system has long been standard practice at both the NIH and the NSF, and some scientists defend it as essential to science as a resource-allocating mechanism.[5] However, the peer-review system has been criticized by some scientists as stifling innovation by encouraging conformity and continuation of proven paths. A 2005 National Research Council report evaluating the postdoc system openly criticized the peer-review system this way:

> The system placed too much emphasis on the number of papers published, too little on whether really important problems were even being tackled. Because requests for grant funds from new investigators were evaluated on the basis of "preliminary results," most funded research became constrained to well-worn research paths—those previously pursued by the new investigators when they were postdoctoral fellows in established laboratories. In short, innovation was the victim of a system that had become much too risk adverse [*sic*].[6]

Whether or not the current peer-review system actually puts American science at risk in the future is not something we have sufficient expertise to judge. Suffice it to say that some prominent scientists have raised this issue as a potential internal threat to American science.[7]

The Science Race in a Global Context

Our study has revealed that American science now faces increasing competition from abroad, manifested in three forms. First, as shown in Chapters 4 and 7, much of American science today relies on the scientific talents of immigrants, in the form of both immigrant scientists and science students in American universities. Second, given the substantial economic resources of the United States, American schoolchildren, as compared with children in other countries, achieve only mediocre scores on international tests of science and mathematics (see Chapter 2). Last, an increasing share of research activities worldwide is now being conducted in other countries, particularly in East Asia (see Chapter 2). If this trend continues, it could lead to the gradual erosion of America's long-standing dominance in world science.

However, as we argued earlier in this book, science has been a globalized enterprise from its inception. Scientists around the world have historically had access to the works of American scientists, especially in basic science, and Americans continually benefit from the work of foreign researchers. Ironically, the very success of American science means that it has improved, both directly and indirectly, the lives of people in other nations, some of which have begun to emulate the success of American science. Those countries, such as China, have achieved rapid improvements in education and research infrastructure and now hope to narrow the science gap between themselves and the United States, especially if they are able to encourage their own overseas students who have enrolled in science programs at U.S. universities to return to their homelands.[8]

Serious competition arising from abroad does not necessarily mean American science is in decline. It may only mean that science today is becoming more globalized. In an age when other countries are catching up, American science will inevitably become less dominant, even when it is not in decline relative to its own past. As more scientists in countries outside the United States and Europe begin to participate in scientific research, the world of science is becoming globalized not only in terms of connectedness but also in terms of active participation in the practice of science. In light of this new world order, American science will surely have difficulty maintaining its historically unchallenged dominance, but this does not mean that science will ultimately be worse off in America because of it.

At a fundamental level, globalization of science is beneficial to science and thus to the whole of humanity. Fuller participation in science by more nations means much greater government investment in research and a much larger science labor force worldwide. Because a scientific discovery needs only to be made once but often benefits all, globalization is certain to speed up scientific advances, for two reasons. First, science as a whole may gain efficiency via complementarity, as scientists in different parts of the world may hold distinct advantages due either to unique natural resources (e.g., access to unusual weather or unusual plants) or to unique intellectual traditions.[9] Second, the sheer expansion of the scientific labor force means more opportunities to produce fruitful scientific results. Hence, globalization of science has the potential to benefit American science as well as American society as a whole.

Of course, globalization of science also presents challenges to American science. One of them is a heightened intensity of competition. With more contenders worldwide competing for recognition, individual American scientists will find it harder in the future to attain top positions in world science. Scientists motivated by the possibility of making pathbreaking discoveries may become particularly discouraged, knowing that more scientists around the world, similarly trained and equipped, are now pursuing the same dream.

A call for better science education and more investment in science in the United States, as represented in the 2007 NAS report, *Rising above the Gathering Storm,* may lead to policy changes that help American science maintain its competiveness and world leadership, which in turn will contribute to the American economy. This call does not need, however, to be based on an alarmist assessment of American science today, although it is often made this way in practice, possibly because federal support for science grew out of past war efforts, including those made during the Cold War after the Sputnik launching.[10] Perhaps due to this unique history, public support for and commitment to science in the United States are strong when the impact of science is couched in terms of national interests.

While most results in basic science are in the public domain and available to all, some scientific knowledge is kept confidential for reasons of either national security or corporate profit, especially in applied science and technology. Thus, if the United States loses its leadership role in basic science and consequently in technological advances, this may, in a knowledge-based economy, have negative effects on our whole society, justifying concerns based on national interests about American science. In the future, competitive countries such as China seem likely to rival—or even surpass—the United States in two areas: adequately equipped scientific laboratories and well-educated scientific personnel. Given the current reliance on the Internet for communication and widespread availability of inexpensive air transportation, physical proximity to existing centers of scientific excellence is becoming less and less crucial. What special assets of American science, then, may still put it at an advantage relative to its foreign competitors? We will answer this question in the next section.

American Science

Science is a social institution. As such, it is interrelated with many other social arenas, such as politics, economics, education, and culture, and must be understood within its social contexts. American science has always been fundamental to American society. In order to understand why, we first need to recognize three defining characteristics of science as a unique social institution: (1) universalism, (2) emphasis on innovation, and (3) devotion to the public good.[11] We believe that these three special features of science fit exceptionally well with certain unique aspects of traditional American culture.

First, universalism has always been a strong norm in America, which was founded as a land of equal opportunity for all.[12] In the absence of a privileged ruling noble class, such as that in Europe, American society has traditionally valued achieved attributes over ascribed characteristics.[13] This cultural emphasis is also known as "individualism." One indication of American individualism is the strong emphasis American culture places on individuals doing their own thinking rather than following that of conventional authorities.[14] For example, America's individualism is manifested in its continual acceptance of immigrants, who gradually assimilate to become full members of American society, both culturally and economically.[15] In terms of cultural norms, no American's achievements or contributions are predetermined by his or her place of birth. Instead, all Americans are supposed to be judged solely on the basis of what they accomplish. This is universalism at work, the same norm that is essential to science.

Of course, American practice falls far short of the American ideal. The processes of educational and occupational attainment are not fully universalistic, either generally or within science.[16] For example, our own analyses in Chapter 7 indicate that, net of their academic performance, the social origins of high school students play a part in determining whether they will receive a bachelor's degree. Within science, too, universalism may be espoused but not fully implemented.[17]

Social norms are nothing but social influences and constraints on individuals' behaviors and beliefs. As such, they are not always followed in practice. However, violations of social norms themselves do not constitute the proof that these norms do not exist. In fact, frequent public outcries against such violations when they occur serve only as indications that social norms do exist and that they do influence and constrain individuals' behaviors and beliefs.[18] The presence of the universalistic norm in America, even if it is not fully implemented, may encourage Americans to pursue science, as it influences the perceptions of young people considering various careers and, thus, affects their career choices. The universalism norm in general may also explain why public confidence in the leadership of science and the prestige of scientists have remained high in recent decades, during which the American scientific labor force has become increasingly composed of women, nonwhites, and immigrants.

Second, American culture strongly emphasizes innovation. From the beginning, the United States was unique not for its appreciation of innovation, which had existed earlier in Europe, but its "belief in the creativity . . . of the ordinary people."[19] In Tocqueville's words, Americans "have swept away the privileges of some of their fellow-creatures which stood in their way, but they have opened the door to universal competition."[20] Indeed, given the universalism discussed above, Lipset proposes that this pressure to innovate may be higher for otherwise disadvantaged social groups:

> Since the emphasis is on individual success in the United States, those individuals or groups who feel themselves handicapped and who seek to resolve their consequent doubts about their personal worth are under strong pressure to "innovate," that is, to use whatever means they can find to gain recognition.[21]

The pressure to be creative is good for science, because science thrives on and rewards innovation. As we described in Chapter 1, American history has been marked by notable technological innovations throughout.

Third, Americans tend to view high levels of achievement as contributing to the public good. One of their unique social beliefs is that "Every Person's Success Improves Society."[22] Note that inequality has been and is a social reality in America, as it is in many other countries.[23] However, Americans have shown relatively little resentment or envy toward that small number of their fellow citizens who enjoy disproportionately large shares of fame and fortune.[24] This tolerance for inequality may be criticized if it contributes to the low degree of income redistribution in the United States and thus exacerbates income inequality. Compared with citizens of other Western countries, Americans are less likely to believe both that it is the responsibility of the government to reduce income inequality and that it is the responsibility of the government to provide a basic income for its members.[25]

One result of the American tolerance for inequality, however, is that instead of the deep class divisions found in European countries, American society is characterized by mass celebration and popular recognition of a small group of highly accomplished citizens.[26] This phenomenon in America can be traced back to several causes.[27] First, in its inception America was a vast frontier land with apparently unlimited resources, and thus competition for success was never viewed as a zero-sum game, as it was in Europe. Second, early American settlers had to cooperate and rely on local communities for survival in a harsh environment, rendering the successes of some helpful to others. Third, high social status was achieved only through hard work in America rather than through either birth or military conquests, as was the case of Europe. Finally, America has always been seen (accurately or inaccurately) as providing equal opportunities to all, whereas in other countries only a few have been selected for recruitment into the elite class.[28] For these reasons, intense competition for social status and recognition has been seen as socially beneficial.[29]

The cultural norm of recognizing individuals' achievements as socially beneficial has been important to the development of science in America, both for the interface between scientists and nonscientists and for dynamics within science. For the former, high social status and recognition have been bestowed on scientists in America with genuine appreciation and admiration.[30] For the latter, scientists have fostered a relatively healthy but highly unequal merit-based reward system that encourages scientific contributions.[31]

The tight fit, as described above, between American values on the one hand and the special features of science on the other has been instrumental in the creation of American science. Although we do not argue that America has a monopoly on the fit between individualistic values and science, we do believe that it has found its fullest expression in this country. Indeed, science was fundamental to America's identity long before America became a world leader in science, as far back as the colonial period, the experiments of Benjamin Franklin, and the earliest years of the American republic.[32] Thus, Samuel Cooper, in a sermon celebrating the commencement of the Constitution in 1780, cited scientific advancements as one happy result of political freedom:

> [Independence] opens to us a free communication with all the world, not only for the improvement of commerce, and the acquisition of wealth, but also for the cultivation of the most useful knowledge. It naturally unfetters and expands the human mind, and prepares it for the impression of the most exalted virtues, as well as the reception of the most important science.[33]

In the long run, a social and cultural environment that is conducive to science may prove to be American science's greatest asset in meeting the challenges of foreign competition. In America's highly competitive but (at least in the ideal) open-to-all and beneficial-to-all environment, science will continue to prosper, as it has in the past. An erosion of such an environment would mean the real decline of American science.

Policy Implications

A book that is motivated by a policy debate is customarily expected to deliver policy recommendations. Although we will do our best to fulfill these expectations, the reader may be disappointed to know that we will not be making any clear-cut recommendations. On a subject as complicated and difficult as science, we feel that any attempt to do so would be presumptuous and foolhardy, especially given the kind of information available to us.

In researching for our book, we have analyzed eighteen large, nationally representative statistical data sets, in addition to a large quantity of published materials and web-based information. These sources of information have been valuable and informative in uncovering certain social realities as we have done in this book, but they do not lend themselves directly to drawing causal conclusions.[34] In other words, our study suffers from data limitations unavoidable in all social science research of the kind we conducted for the book.

Some of the realities we have uncovered, however, do seem to have important ramifications for policymakers. In acknowledging these here, we offer them more as points for further attention in the science policy debate than as policy recommendations.

First, scientists' earnings have stagnated in the post-1960 decades against a generally increasing trend in earnings for the high-educated labor force during the same period. This decline has lowered the overall cost of conducting science in the United States, but it also makes science careers less attractive than they would otherwise have been. What to do about the situation is, however, less clear. Enhancing the demand for scientists by increasing government spending would boost the earnings of scientists overall but would also raise the cost of doing science.[35] Furthermore, if the increased spending is mainly in the form of research grants that would add to the budgets of existing research labs without leading to more permanent research positions, the increase may translate in practice to the more temporary junior positions considered "glut" by some science observers.[36] Clearly, the solution is not just about spending more money: choices about how to spend it have real consequences. Spending in one area necessarily means reducing spending in other areas, or increasing debt. Thus, any decision to increase funding in science will come at the cost of other benefits that are also valued by the public, such as general education or health care—items that receive more support for increased funding than does science. However, inadequate spending in science would speed up the decline of America's dominance in world science.

Second, academic science may have cause for concern, but the solution is not clear. At the same time that the high risk associated with academic careers in science may deter risk-averse young people from pursuing this career, the funding process in science has been criticized for rewarding risk too little, stifling innovation. Despite the criticisms, the practice of the postdoc system will most likely continue into the future. As a 2005 National Research Council report already recommended, funding agencies such as the NIH should devise new guidelines and programs that would improve the lives of postdocs and encourage them to do independent research.[37] It would also be beneficial for these funding agencies to work closely with universities on some of the issues.

Third, international competition is real, as the science world is becoming progressively more globalized. However, it would be shortsighted for American science to be closed to outsiders, as science in the United States benefits from, indeed depends on, globalization. Scientific advances in other countries should help American science, as U.S. scientists will not only build on concrete contributions made by foreign scientists but also be stimulated by competition with them. Furthermore, much of the science that takes place today on American soil is practiced by scientists who have immigrated to the United States. These immigrants make important contributions to American science, just as earlier generations of immigrant scientists did.[38] It is appropriate, however, to consider two possible risks for American science in overrelying on immigrant scientists: (1) the supply of immigrant scientists may decline or even stop altogether, perhaps unexpectedly, in the future; and (2) the supply of immigrant scientists may lower the overall earnings of scientists and thus reduce the attractiveness of science careers to native-born Americans.

Finally, as a policy matter, science education should not be detached from general education. The twentieth century was the American century mainly because America benefited from its huge investment in education—human capital.[39] Indeed, there is an intrinsic, dynamic relationship between science and general education: not only is an improvement in education a precondition for training scientists, technological advances made by scientists change the direction of the American economy, which, in turn, demands better-educated workers. There are some signs that America's investment in higher education has slowed down in recent decades.[40] One piece of empirical evidence consistent with the underinvestment argument is a corresponding increase since the 1980s in earnings for highly educated workers.[41] To enhance general education, the United States would need to meet at least three sets of challenges ahead. First, it needs to narrow the gap in accessing education, especially in the early years, by race and family socioeconomic background.[42] Second, further investment in high-quality colleges and universities is needed to accommodate increasing demands.[43] Third, the academic mission for knowledge acquisition and training in American institutions of higher education needs to be strengthened.[44]

America is a great country, for both the gifted and the ordinary, native-born Americans and immigrants, scientists and nonscientists. In its relatively short history, America has faced serious challenges, such as the Revolutionary War, the Civil War, World War I, the Great Depression, World War II, and the Cold War, but it has surfaced each time from these periods of difficulty with greatly improved strengths. One of this nation's greatest assets has been the close affinity between American science and American society at large. In the long run, this invaluable asset will prove instrumental in keeping science strong in America, giving us good reason to remain optimistic about the future of American science. While America may not always remain dominant in a world that is becoming increasingly interconnected scientifically, this does not mean that science in America is doomed to mediocrity. Loss of dominance does not mean decline. All current signs indicate that American science can still remain a leader of world science for many years to come.

Census and American Community Survey Data

Tables and figures based on census data are constructed using the 1 percent Public Use Microdata Samples (PUMS) of the decennial U.S. censuses in 1960, 1970, 1980, 1990, and 2000. Similar tables and figures for 2006 through 2008 are based on the 2006–2008 three-year American Community Survey Public Use Microdata Sample, a subsample of cases from the American Community Survey (ACS), an annual survey of 5 percent samples of PUMAs (non-overlapping state partitions each with a population of about 100,000 residents). In all analyses, we restrict the sample to employed members of the civilian labor force, aged twenty-five to fifty-nine. In 1960, scientists were classified as working in biology, engineering, mathematics, or the physical sciences. In 1970, 1980, 1990, 2000, and the ACS, computer science was another possible classification. In all years, we require that those classified as scientists hold at least a bachelor's degree. The complete occupation-coding scheme can be found in Codebook A at http://www.yuxie.com.

For 1970, 1980, and 1990, the census data include separate codes for academics working in different disciplines. Thus, in every data set except 2000 and the 2006–2008 ACS, it is possible to separate the population of scientists into academic scientists and nonacademic scientists. In the 2000 PUMS and the 2006–2008 ACS, however, the data contain only a single code for all academics. We used the 1990 data to estimate the fraction of scientists within each field who were academics. We then allocated academic scientists in 2000 and 2006–2008 to these fields in such a way as to keep the share of academics within the field constant across years 1990, 2000, and 2006–2008.

For example, in the 1990 data there were 86 biological academic scientists and 1,132 biological nonacademic scientists. Thus, the share of academic scientists in the field of biology was 7.1 percent (i.e., 86 out of 86 + 1,132). We held this share constant for 2000 and 2006–2008. For biology, for example, this procedure yielded 137 academic scientists among a total of 1,936 biological scientists in 2000. This same procedure was used for each scientific discipline.

In order to select academics for imputation in scientific fields in 2000 and the 2006–2008 ACS, we used five separate logit regression models, one for each field (physical science, biological science, mathematical science, engineering, and computer science) to predict the probability that an academic would be a scientist in each field based on age, sex, and degree level in the 1990 data set. On the basis of the estimated logit regression models, we predicted the probability that an academic in 2000 or 2006–2008 would be in each scientific field and used this probability as a weight in randomly selecting the appropriate number of academics for each scientific field. Individuals who were assigned to multiple fields by this method were then assigned to whichever field had the most academic scientists.

Table 4.2, which compares the mean income of male scientists to that of members of selected other occupations, makes use of a subsample of the data and uses additional information. The results are broken down by the respondent's most advanced degree and specific scientific fields (biological sciences, physical sciences, etc.). These results for females are available in Table D4.3. We focus on the results for men in the main analysis because the smaller number of female scientists, especially in the earlier years, makes our estimates for women much less precise.

In the 1960, 1970, and 1980 data sets, it was necessary to construct the respondent's highest degree obtained from data on the number of years of education the respondent completed, whereas in the 1990, 2000, and 2006–2008 data sets, respondents were specifically asked to name the highest degree they had obtained. There is no perfect method of keeping degree coding consistent in the earlier years. Coding respondents with seventeen years of education as having a bachelor's degree would underestimate the number of respondents with an advanced degree, whereas coding respondents with seventeen years of education as having an advanced degree would overestimate the number of respondents with an advanced degree.[1] We coded respondents with seventeen finished years of education in two ways—as having a bachelor's degree and as having an advanced degree—but find that our results are almost identical, no matter which coding scheme we use. We chose to report results based on the procedure of coding respondents with sixteen finished years of education as having a bachelor's degree and respondents with seventeen or more years of finished education as having an advanced degree.

While the 1980 data allowed us to construct a variable for highest degree obtained that differentiated between master's and doctoral degrees, the 1960 and 1970 data did not. To impute earnings differences between scientists and other professions at the master's and doctoral levels in the 1960 and 1970 data, we used a decomposition method based on two assumptions: (1) constant proportions of scientists with master's degrees and with doctoral degrees (P_{mK} and P_{pK}, respectively, which sum to one) for each profession (K) among advanced degree holders, and (2) the constant earnings premium of a doctoral degree over a master's degree for each profession (controlling for age and work hours per week). Due to the constancy assumptions, we can estimate unknown quantities in 1960 and 1970 from the 1980 data. To these pieces of information, we further add γ_{YK}, the estimated difference in logged earnings between scientists and members in each profession (K) among those with advanced degrees (controlling for age and work hours per week) in the year (Y) for which we are imputing. The following two equations describe the decompositional relationships (separately by sex):

$$\gamma_{YK} = P_{mK} * \lambda_{mYK} + P_{pK} * \lambda_{pYK}$$

$$\lambda_{pYK} = premium(p_K) + \lambda_{mYK}.$$

Given the two equations, we can solve for two unknowns, highlighted by bold fonts in the two equations. Note that $\gamma_{YK} = 0$ for basic scientists as a special case.

For degree levels for which we did not have to impute earnings differences (bachelor's degrees in all data sets and master's and doctoral degrees in the 1980, 1990, 2000, and 2006–2008 data sets), we calculated earnings ratios relative to scientists by running ordinary least squares regressions for each degree level in each data set, separated by sex, of the log of earnings on a set of profession dummy variables (with scientists as the reference category) controlling for age and a linear spline function of hours of work per week (divided into 35–40 hours, 41–50 hours, and 51+ hours categories). Thus, in each case the earnings ratio of scientists to another profession was the exponential coefficient for that profession in the regression.

Note that all models concerning hours and earnings are restricted to full-time workers ages thirty-five to forty-five. Full-time work is defined as involving weekly work hours greater than 35, annual work weeks greater than 50, and earnings greater than a total equal to $5,000 in 2000 dollars using the Consumer Price Index.[2]

For the 2006–2008 ACS, we used earnings adjustment values included in the data set (which account for the relative value of the dollar in the month and year of the interview) to standardize earnings to 2008 dollars. For all data sets, to calculate the growth rate of real earnings by profession, we adjusted earnings for each year to be comparable to 2000 dollars using the Consumer Price Index.

NCES Survey Data

Our data come from three panel surveys of American students conducted by the National Education Longitudinal Studies Program at the National Center for Education Statistics (NCES). Each study follows a specific cohort of students. In the first study, the National Longitudinal Study of the Class of 1972 (NLS-72), which began in the spring of 1972, students who were high school seniors at that time were interviewed. The original sample was randomly drawn from a probability sample of public and private high schools in the fifty states and the District of Columbia. Follow-up surveys were conducted in 1973, 1974, 1976, 1979, and 1986.

The second study, High School and Beyond (HS&B), began in the spring of 1980 with two cohorts of students, one then in their sophomore year and one in their senior year. The original sample was selected using a two-stage probability sampling design: the first stage sampled public and private schools, and the second stage selected students within schools. Follow-up surveys were conducted of both cohorts in 1982, 1984, and 1986 and of the sophomore cohort in 1992. In order to maintain equal spacing between our three groups, we use the information from the sophomore cohort, which becomes the graduating class of 1982. In order to maintain comparability with the other samples, sophomores who were not enrolled as seniors in 1982 are deleted from the sample.

The third study, the National Education Longitudinal Study of 1988 (NELS), began in 1988 with a sample of eighth-grade students. The original sample of students was selected by a two-stage probability sampling design from public and private high schools in the United States. Follow-up surveys were conducted in 1990, 1992, 1994, and 2000. The sample was freshened in 1990 and 1992 to ensure that it would be representative of students who were sophomores and seniors in 1990 and 1992, respectively. We make use of the freshened 1992 sample in order to have a sample that is representative of high school seniors in 1992.

We use data from these three surveys to measure both the expectation of receiving a bachelor's degree in science among high school seniors and the attainment of a bachelor's degree in a scientific field. Separate analytic samples were constructed for the two analyses. Although the independent variables for the two analyses are identical, we do not require that observations contributing to the analysis of expectations have valid data on educational attainment, and vice versa. The NLS-72 cohort includes 2,967 males and 3,429 females in the expectation sample and 3,769 males and 4,300 females in the attainment sample. The HS&B cohort includes 3,190 males and 3,610 females in the expectation sample and 3,360 males and 3,790 females in the attainment sample. Finally, the NELS cohort includes 3,650 males and 4,060 females in the expectation sample and 4,120 males and 4,560 females in the attainment sample.[1]

We invoke a two-step conceptualization of scientific training, which focuses on the level of educational attainment in the first step and the field of study in the second step. Corresponding to this orientation, in the analysis of students' aspirations, we focus on whether the student expects to receive a bachelor's degree and, conditional on expecting a bachelor's degree, whether the student expects to major in a scientific field. Likewise, in our analysis of educational outcomes, our key outcome measures are whether the individual receives a bachelor's degree and, conditional on receiving a bachelor's degree, whether it was in a scientific field. Thus, we have four outcome variables, two in the analysis of expectations and two in the analysis of attainment. They are measured as follows:

EXPECTATION OF BACHELOR'S DEGREE. In the 1982 and 1992 cohorts, students were asked about their expected levels of educational attainment. Those who selected a bachelor's, graduate, or professional degree were coded as expecting to receive at least a bachelor's degree. For the 1972 cohort, the educational aspirations question does not ask about degree attainment but about attendance at institutions. For this data set, we approximate the set of students who anticipate receiving bachelor's degrees with the set of students who report expecting to attend a four-year college or university or to attend graduate or professional school after graduating from college.

EXPECTATION OF A SCIENTIFIC DEGREE. Students who indicated that they expected to pursue postsecondary education were asked to select their most likely field of study from a list. Among those who expected to receive at least a bachelor's degree, students were coded as expecting to receive a scientific degree if they selected biological sciences, computer and information sciences, mathematics, engineering, or physical sciences. Some analyses also report results disaggregated into these five subfields.

BACHELOR'S DEGREE. Receipt of a bachelor's degree is measured by a binary variable that is set to one if the individual achieves a bachelor's degree prior to the end of the calendar year eight years after the calendar year of his/her high school graduation.[2]

SCIENTIFIC DEGREE. Receipt of a scientific degree is measured by a binary variable that is set to one if the individual completes a bachelor's degree in a scientific field. Individuals who receive multiple bachelor's degrees are considered to have received a bachelor's degree in science if their primary field of study for any of the bachelor's degrees was a scientific one. We consider degrees received in physical science, life science, engineering, math and statistics, or computer science to be scientific degrees. Some analyses report results disaggregated by these five scientific subfields.

Degrees received in agricultural sciences are considered as a subset of life science degrees. Degrees received in the social sciences are not considered science degrees.[3] A complete list of the fields considered scientific for each study, as well as the subfield of science in which they are classified, is given in Codebooks C, D, and E, which are posted online at http://www.yuxie.com.

Our analytic technique is sequential logistic regression models for our two primary outcomes of interest: whether the student achieved a bachelor's degree and, conditional on bachelor's degree attainment, whether he or she received a bachelor's degree in a scientific field. We perform analogous models for expectation of a bachelor's degree and of a scientific bachelor's degree. For each outcome, we consider two separate models, the first of which controls only for sociodemographic characteristics of the student, which we assume to be exogenous. In the second model, we also control for measures of the student's academic achievement in high school. The first model yields the total effects of the sociodemographic characteristics, while the second model estimates the direct effects of the sociodemographic characteristics net of those that operate through academic achievement. Given the emphasis on objectively measurable criteria in science, we expect that the influences of family background on attainment of science degrees should be mostly indirect through academic performance and attainment of degrees overall.

In analyzing the data, we have attempted to make the measurement of key variables as comparable as possible across the data sets. Whenever possible, we use students' demographic information supplied in the questionnaire completed during their senior year of high school. When values for demographic variables are missing for a variable in the senior year survey but available for other waves of the survey, we fill in the missing values with information from other waves. Academic performance was assessed through a series of tests administered by the surveys. For these variables, we make use only of the scores received in the senior year. The independent variables that we use in our regression models include the following:

SEX. Students' self-reported sex is used to classify students as either male or female.

RACE. Students' self-reported race is used to classify students as white, Asian, Hispanic, or black. Because of the small number of respondents of other races, particularly in earlier cohorts, individuals who identified themselves as members of a race other than these four major groups were dropped from the sample.

MATERNAL EDUCATION. The education of the respondent's mother is categorized into four categories: less than high school graduation, high school graduation but no college, some college but no bachelor's degree, or a bachelor's degree or higher.

FAMILY STRUCTURE. A dummy variable is created to indicate whether the respondent was living with both parents (biological or adoptive) during the spring of his or her senior year.

FAMILY INCOME. For each cohort, respondents were provided with a list of family income ranges and were asked to indicate the category of their families' incomes during the respondent's senior year of high school. The number of categories provided varied across surveys, with ten categories provided in NLS-72, eight in HS&B, and fifteen in NELS. We construct a new relative family income variable that is consistent across the data sets, measuring, approximately, the percentile of the respondent's family in terms of household income. We use single imputation based on other background information (race, number of siblings, respondent's maternal education, and the family structure) for respondents who did not provide a valid response to the income question.[4]

ACADEMIC ACHIEVEMENT. For each respondent, we create an achievement score in reading and an achievement score in math. For cohorts that include multiple measures in reading or math, the measures were averaged. The reading and math scores were then transformed into percentile measures in each subject.

The math percentile score enters the model directly. However, to capture the respondent's comparative advantage in science careers, we construct a second variable, *premium,* which is the difference between the student's percentile scores in math and in reading. A positive value of premium indicates that the student scored in a higher percentile in math than in reading.

All descriptive statistics and analyses are weighted to best approximate a representative sample of high school seniors in 1972, 1982, and 1992 for the NLS-72, HS&B, and NELS, respectively.

APPENDIX C

NES, NSRCG, and IPEDS Data

In Chapters 7 and 8, we describe the transitions to graduate programs and to scientific occupations of science degree recipients over the past three decades. Our data for these analyses come from two studies conducted by the National Science Foundation (NSF): the New Entrants Survey and the National Survey of Recent College Graduates.

The New Entrants Surveys (NES), also known as "Surveys of Recent Science and Engineering Graduates," were a series of cross-sectional studies that collected data on personal, educational, and occupational characteristics of new entrants into the labor market. The surveys were conducted for the NSF every two years from 1976 to 1988. In each survey year, the sample was drawn from all U.S. citizens and permanent residents who graduated with a bachelor's or master's degree in scientific fields within the previous two academic years. Each NES sample was determined through a two-stage probability sampling procedure. At the first stage, degree-granting institutions were selected in proportion to their relative contributions to the entire scientific degree production process. Next, individual graduates were selected within each school. Those from smaller disciplines had greater probabilities of being chosen.

The National Survey of Recent College Graduates (NSRCG) is structurally similar to the NES. It was carried out biennially for the NSF to collect information on recent college graduates in the United States, with a focus on their transitions to further studies or occupations. In each survey, the population consisted of all individuals who received a bachelor's or master's degree in science, engineering, or health over the two academic years prior to the survey year. Like NES, the NSRCG adopted a two-stage sampling scheme. First, a probability sample of institutions was chosen in proportion to size from a list provided by the National Center for Education Statistics (NCES). Adjustment was made so that schools with relatively high proportions of graduates in minority groups had higher probabilities of being selected. At the second stage, eligible degree recipients were selected with differential sampling rates for fields of study that enabled over- or undersampling.

149

Our analysis uses six years of data from NES and two recent years from NSRCG. All analyses using NES and NSRCG data were weighted. Due to data limitation, we have to leave a nontrivial gap between the years 1988 and 2003. All samples are restricted to bachelor's or master's degree recipients in four scientific subfields (biological sciences, physical sciences, mathematical sciences, and engineering) who graduated within the previous two academic years. For example, the NES 1984 sample consists of individuals who earned their degrees in 1982 and 1983. The only exception is NSRCG 2006, which by design includes graduates over the previous three years. The subfields have been made as comparable as possible across the studies. The detailed coding is documented in Codebooks F and G, provided online at http://www.yuxie.com. After restriction, the analytical sample sizes are as follows:

NES Bachelor's Degree Recipients Samples

1976: 5,234
1980: 5,548
1982: 6,518
1984: 7,882
1986: 7,694
1988: 7,865

NSRCG Bachelor's Degree Recipients Samples

2003: 4,907
2006: 7,479

NES Master's Degree Recipients Samples

1976: 587
1980: 2,087
1982: 2,513
1984: 3,262
1986: 3,420
1988: 3,691

NSRCG Master's Degree Recipients Samples

2003: 2,040
2006: 3,187

Two types of transition rates are calculated using the above samples. As illustrated in Figure 7.3, we conceptualize three outcome statuses after one obtains a bachelor's or master's degree in science: further education, working, and other. In Chapter 7 (Tables 7.2–7.3 and Appendix Tables D7.6–7.7), we report the transitions from scientific degrees to graduate programs. The transition considered in Chapter 8 pertains to the utilization of science degrees. Table 8.1 and Appendix Table D8.1 contain the observed rates of going from a science degree to a science job, conditional on working. In other words, among all working individuals who earned a bachelor's or master's degree in science or engineering within the previous two years, we ask how many were employed in scientific occupations. In addition to our own analyses, we also borrow from the 2006 report of Survey of Earned Doctorates (a census of all "research doctorate recipients" in the United States) a roughly parallel utilization time series at doctoral level (Table 8.2). Note that the doctorate data are different in field categorization as well as sample restriction and therefore cannot be directly compared with the results concerning lower-level degrees.

IPEDS

To aid discussions in Chapter 7, three degree production trends are constructed at bachelor's, master's, and doctoral levels (Figure 7.1). To maintain comparability, the degrees are restricted to four scientific subfields—biological, physical, mathematical, and engineering (cf. Codebook B). The data are retrieved in aggregate form from the online tabulation system WebCASPAR at http://webcaspar.nsf.gov, which covers the years 1966 to 2008. The data source is the Integrated Postsecondary Education Data System (IPEDS) Completions Survey administered by the NCES, of the Department of Education. From the same source, we also calculated two thirty-year time series of gender and immigrant-specific science degree production at all three levels of degrees (Figure 7.2).

Detailed Statistical Tables

This appendix provides additional tables that are intended for readers who are interested in detailed statistical results.

Table D4.1. Median gender earnings ratio of U.S. scientists, 1960–2008

	1960	1970	1980	1990	2000	2006–2008
Gender						
Male	9,500	15,300	28,005	47,000	67,000	88,614
Female	5,950	10,800	20,285	38,000	57,250	75,924
Female/male ratio	0.63	0.71	0.72	0.81	0.85	0.86

Sources: U.S. Census 1960–2000; American Community Survey 2006–2008.

Table D4.2. Median racial earnings ratio of U.S. scientists, 1960–2008

	1960	1970	1980	1990	2000	2006–2008
Race/ethnicity						
White (W)	9,500	15,200	28,005	45,300	65,000	86,563
Black or Hispanic (BH)	7,600	14,000	24,005	41,000	58,000	75,924
BH/W ratio	0.80	0.92	0.86	0.91	0.89	0.88

Sources: U.S. Census 1960–2000; American Community Survey 2006–2008.

Table D4.3. Estimated ratios in earnings between professionals and scientists, by degree and decade (female workers only)

	1960	1970	1980	1990	2000	2006–2008
Bachelor's degree						
Scientists (bio., math, phys.)	*1.00*	*1.00*	*1.00*	*1.00*	1.00	1.00
Engineers	*1.53*	*1.16*	*0.97*	1.10	1.32	1.18
Computer scientists	N/A	1.13	*1.13*	1.11	1.26	1.13
Nurses	*0.91*	0.76	0.91	1.00	1.01	0.99
Teachers	0.83	0.66	0.73	0.74	0.70	0.64
Social scientists	*0.73*	*1.04*	*0.83*	*1.02*	*1.19*	1.08
Master's degree						
Scientists (bio., math, phys.)	*1.00*	*1.00*	*1.00*	*1.00*	1.00	1.00
Engineers	N/A	*0.97*	*1.04*	*1.38*	1.23	1.21
Computer scientists	N/A	*0.90*	*0.98*	1.21	1.30	1.16
Nurses	*0.89*	*0.83*	0.91	1.20	1.10	1.08
Teachers	0.86	0.71	0.77	0.91	0.81	0.75
Social scientists	*1.04*	*1.19*	*1.03*	1.04	0.97	1.03
Doctorate (PhD & professional)						
Scientists (bio., math, phys.)	*1.00*	*1.00*	*1.00*	*1.00*	1.00	1.00
Engineers	N/A	*1.27*	*1.20*	*1.53*	*1.69*	*1.23*
Computer scientists	N/A	*1.15*	*1.14*	*1.08*	*1.24*	1.17
Social scientists	*1.10*	*1.26*	*1.07*	*1.19*	*1.08*	0.99
Doctors	*1.75*	*1.74*	*2.00*	1.76	1.92	1.62
Lawyers	*0.60*	*1.04*	*1.16*	1.40	1.44	1.34

Sources: U.S. Census 1960–2000; American Community Survey 2006–2008.
Notes: Analysis is restricted to full-time, full-year workers.
Ratios are computed using scientists' earnings as the benchmark.
Estimates based on fewer than 100 cases are presented in italics.

Table D5.1. Responses to scientific literacy questions, by religious identification (percent)

	Protestant (*n*=986)	Catholic (*n*=402)	Other (*n*=45)	None (*n*=325)
The center of the earth is very hot. (True)	78.3	77.8	74.7	88.6
All radioactivity is man-made. (False)	68.4	68.4	69.2	76.1
It is the father's gene that decides whether the baby is a boy or a girl. (True)	66.6	66.6	67.7	53.8
Lasers work by focusing sound waves. (False)	42.5	42.3	45.6	55.6
Electrons are smaller than atoms. (True)	51.6	54.4	52.4	58.4
Antibiotics kill viruses as well as bacteria. (False)	54.5	55.6	75.4	56.9
The universe began with a huge explosion. (True)	24.8	37.2	49.0	43.9
The continents on which we live have been moving their locations for millions of years and will continue to move in the future. (True)	75.7	82.6	68.0	86.6
Human beings, as we know them today, developed from earlier species of animals. (True)	29.8	51.5	53.8	64.2
Does the earth go around the sun, or does the sun go around the earth? (Earth around the sun)	72.0	75.5	79.4	81.0
[Among those who answered that earth goes around the sun] How long does it take the earth to go around the sun: one day, one month, or one year? (One year)	70.6	68.5	72.2	79.8

Source: General Social Survey 2006 (Smith et al. 2011).
Note: All percentages are calculated with weighted sample sizes.

Table D6.1. Transition rates from expectation to attainment, by gender and cohort (percent)

	Attainment							
	Male				Female			
Expectation	No bachelor's	Nonscience bachelor's	Science bachelor's	Total	No bachelor's	Nonscience bachelor's	Science bachelor's	Total
1972 cohort (NLS-72)								
No bachelor's	92.2	6.0	1.8	100	94.6	5.0	0.4	100
Nonscience bachelor's	45.1	48.4	6.5	100	41.4	55.5	3.1	100
Science bachelor's	36.7	27.2	36.2	100	37.2	36.3	26.5	100
Total	71.7	20.2	8.0	100	76.0	21.6	2.5	100
				(N=3,566)				(N=4,174)
1982 cohort (HS&B)								
No bachelor's	92.6	5.9	1.5	100	93.5	5.6	1.0	100
Nonscience bachelor's	40.6	51.5	7.9	100	42.1	53.2	4.6	100
Science bachelor's	40.3	21.6	38.1	100	37.5	34.1	28.5	100
Total	73.5	18.0	8.5	100	73.6	22.8	3.6	100
				(N=2,940)				(N=3,380)
1992 cohort (NELS)								
No bachelor's	93.0	6.0	1.0	100	94.4	4.6	1.0	100
Nonscience bachelor's	51.4	41.4	7.2	100	40.0	54.8	5.2	100
Science bachelor's	45.2	19.3	35.6	100	34.3	35.9	29.8	100
Total	69.3	22.0	8.7	100	63.0	32.2	4.9	100
				(N=3,910)				(N=4,370)

Sources: NLS 1972; HS&B 1980; NELS 1988 (see Appendix B).

Notes: All percentages are calculated using weighted sample sizes.
Sample sizes for the HS&B and NELS cohorts are rounded to the nearest ten, in accordance with Institute of Education Statistics policy.

Table D6.2. Descriptive statistics, by gender and cohort (percent)

	Male			Female		
	1972 cohort (NLS-72)	1982 cohort (HS&B)	1992 cohort (NELS)	1972 cohort (NLS-72)	1982 cohort (HS&B)	1992 cohort (NELS)
Race						
White	92.8	75.7	75.6	90.4	76.4	73.4
Asian	0.9	1.1	4.2	1.0	1.0	3.7
Hispanic	2.2	13.2	7.9	2.3	11.2	10.1
Black	4.2	10.0	12.3	6.3	11.5	12.8
Mother's education						
Less than high school	19.4	14.9	11.1	25.2	19.1	14.1
High school graduate	55.1	55.6	49.6	52.4	52.3	51.4
Some college	11.2	13.2	13.3	10.0	14.9	11.3
College graduate or higher	14.3	16.3	26.0	12.4	13.7	23.2
Family structure						
Two-parent	78.5	77.0	66.1	77.3	73.8	66.3
Other	21.5	23.0	33.9	22.7	26.2	33.7
Sample size	2,967	3,190	3,650	3,429	3,610	4,060

Sources: NLS 1972; HS&B 1980; NELS 1988 (see Appendix B).
Note: Sample sizes for the HS&B and NELS cohorts are rounded to the nearest ten, in accordance with Institute of Education Statistics policy.

Table D6.3. Percentage of respondents expecting a bachelor's degree, by demographic subgroups

	Male			Female		
	1972 cohort (NLS-72)	1982 cohort (HS&B)	1992 cohort (NELS)	1972 cohort (NLS-72)	1982 cohort (HS&B)	1992 cohort (NELS)
Race						
White	56.9	48.8	65.8	45.3	47.1	71.7
Asian	68.1	71.0	74.4	78.7	68.1	80.1
Hispanic	46.3	26.6	65.6	36.0	30.4	55.2
Black	52.7	33.1	61.5	53.8	48.3	66.2
Family income						
Quartile 1	42.0	31.4	48.1	34.8	32.4	55.2
Quartile 2	49.6	41.9	63.5	38.1	41.8	65.4
Quartile 3	59.6	49.4	72.9	49.9	53.9	79.7
Quartile 4	74.7	64.5	88.2	66.6	70.7	95.8
Mother's education						
Less than high school	36.7	22.8	40.0	31.1	25.8	46.8
High school graduate	53.4	38.2	59.9	42.8	40.5	63.6
Some college	77.3	61.2	71.4	67.7	61.3	80.1
College graduate or higher	79.6	72.4	84.7	72.3	75.6	91.8
Family structure						
Two-parent	59.8	46.4	70.5	48.2	46.9	72.3
Other	45.0	38.5	56.2	38.7	41.9	64.4

(continued)

Table D6.3 *(continued)*

	Male			Female		
	1972 cohort (NLS-72)	1982 cohort (HS&B)	1992 cohort (NELS)	1972 cohort (NLS-72)	1982 cohort (HS&B)	1992 cohort (NELS)
Math score						
Quartile 1	25.1	12.3	36.6	22.0	21.1	44.3
Quartile 2	44.9	28.1	57.9	40.5	36.8	64.3
Quartile 3	68.1	48.0	74.3	63.6	54.0	77.6
Quartile 4	86.3	80.3	92.9	82.6	81.9	95.2
Premium						
Quartile 1	50.5	36.1	56.3	42.2	43.1	64.7
Quartile 2	58.0	48.2	69.2	51.4	48.3	71.5
Quartile 3	58.9	53.6	64.5	44.9	48.3	69.4
Quartile 4	56.8	40.4	70.3	46.5	42.7	75.6
Sample size	2,967	3,190	3,650	3,429	3,610	4,060

Sources: NLS 1972; HS&B 1980; NELS 1988 (see Appendix B).

Notes: Premium is the difference between the standardized math score and the standardized verbal score.

Sample sizes for the HS&B and NELS cohorts are rounded to the nearest ten, in accordance with Institute of Education Statistics policy.

Table D6.4a. Logistic regression results predicting expectation of a bachelor's degree, for males

| | 1972 cohort (NLS-72) | | | | 1982 cohort (HS&B) | | | | 1992 cohort (NELS) | | | |
| | Simple model | | Full model | | Simple model | | Full model | | Simple model | | Full model | |
Variable	Coef.	Robust SE	Coef.	Robust SE	Coef.	Robust SE	Coef.	Robust SE	Coef.	Robust SE	Coef.	Robust SE
Race (white=excluded)												
Asian	0.46	(0.42)	0.24	(0.41)	1.01	(0.32)**	0.92	(0.35)**	0.37	(0.35)	0.27	(0.37)
Hispanic	0.20	(0.26)	0.97	(0.30)**	−0.68	(0.14)***	0.08	(0.16)	0.64	(0.20)**	1.05	(0.22)***
Black	0.44	(0.22)*	1.26	(0.23)***	−0.39	(0.17)*	0.67	(0.21)**	0.20	(0.28)	0.99	(0.33)**
Family income (Quartile 1=excluded)												
Quartile 2	0.18	(0.12)	0.05	(0.14)	0.28	(0.13)*	0.28	(0.14)	0.42	(0.17)*	0.27	(0.19)
Quartile 3	0.40	(0.13)**	0.33	(0.15)*	0.42	(0.15)**	0.35	(0.17)*	0.71	(0.23)**	0.71	(0.23)**
Quartile 4	0.91	(0.13)***	0.68	(0.15)***	0.84	(0.15)***	0.77	(0.17)***	1.46	(0.27)***	1.02	(0.28)***
Mother's education (less than high school=excluded)												
High school graduate	0.51	(0.11)***	0.36	(0.13)**	0.56	(0.15)***	0.44	(0.17)*	0.65	(0.19)**	0.61	(0.19)**
Some college	1.44	(0.18)***	1.16	(0.20)***	1.38	(0.19)***	1.11	(0.23)***	1.09	(0.30)***	0.80	(0.32)*
College graduate or higher	1.52	(0.18)***	1.31	(0.19)***	1.83	(0.19)***	1.26	(0.21)***	1.68	(0.23)***	1.27	(0.23)***
Family structure (other=excluded)												
Two-parent	0.44	(0.11)***	0.39	(0.13)**	0.12	(0.13)	0.08	(0.15)	0.37	(0.17)*	0.25	(0.17)

(continued)

Table D6.4a (*continued*)

Variable	1972 cohort (NLS-72) Simple model Coef.	Robust SE	Full model Coef.	Robust SE	1982 cohort (HS&B) Simple model Coef.	Robust SE	Full model Coef.	Robust SE	1992 cohort (NELS) Simple model Coef.	Robust SE	Full model Coef.	Robust SE
Math score (Quartile 1 = excluded)												
Quartile 2			1.07	(0.15)***			1.23	(0.19)***			0.98	(0.22)***
Quartile 3			2.08	(0.15)***			2.19	(0.19)***			1.71	(0.20)***
Quartile 4			3.20	(0.18)***			3.73	(0.21)***			3.14	(0.26)***
Premium (Quartile 1 = excluded)												
Quartile 2			−0.29	(0.16)			−0.36	(0.17)*			0.05	(0.24)
Quartile 3			−0.37	(0.15)*			−0.34	(0.16)*			−0.16	(0.23)
Quartile 4			−0.87	(0.15)***			−1.24	(0.17)***			−0.23	(0.22)
Constant	−1.10	(0.14)***	−1.93	(0.20)***	−1.34	(0.18)***	−2.74	(0.25)***	−1.01	(0.22)***	−1.98	(0.27)***
Model χ^2	257.6		545.3		263.6		528.7		141.7		361.4	
Degrees of freedom	10		16		10		16		10		16	
N	2,967		2,967		3,190		3,190		3,650		3,650	

Sources: NLS 1972; HS&B 1980; NELS 1988 (see Appendix B).

Notes: Premium is the difference between the standardized math score and the standardized verbal score. Sample sizes for the HS&B and NELS cohorts are rounded to the nearest ten, in accordance with Institute of Education Statistics policy.

*$p<.05$. **$p<.01$. ***$p<.001$.

Table D6.4b. Logistic regression results predicting expectation of a bachelor's degree, for females

	1972 cohort (NLS-72)				1982 cohort (HS&B)				1992 cohort (NELS)			
	Simple model		Full model		Simple model		Full model		Simple model		Full model	
Variable	Coef.	Robust SE	Coef.	Robust SE	Coef.	Robust SE	Coef.	Robust SE	Coef.	Robust SE	Coef.	Robust SE
Race (white=excluded)												
Asian	1.62	(0.61)**	1.57	(0.77)*	0.88	(0.31)**	0.94	(0.33)**	0.34	(0.33)	0.33	(0.34)
Hispanic	0.15	(0.24)	0.86	(0.27)**	-0.37	(0.15)*	0.34	(0.17)*	-0.09	(0.24)	0.23	(0.25)
Black	0.85	(0.16)***	1.76	(0.17)***	0.39	(0.17)*	1.33	(0.18)***	0.23	(0.36)	0.57	(0.45)
Family income (Quartile 1=excluded)												
Quartile 2	0.15	(0.11)	0.05	(0.12)	0.24	(0.11)*	0.16	(0.13)	0.23	(0.18)	0.10	(0.19)
Quartile 3	0.45	(0.12)***	0.47	(0.13)***	0.67	(0.14)***	0.55	(0.15)***	0.67	(0.18)***	0.48	(0.19)*
Quartile 4	0.97	(0.13)***	0.88	(0.14)***	1.24	(0.15)***	1.15	(0.17)***	2.21	(0.25)***	1.86	(0.26)***
Mother's education (less than high school=excluded)												
High school graduate	0.37	(0.10)***	0.19	(0.11)	0.52	(0.13)***	0.26	(0.14)	0.47	(0.17)**	0.26	(0.19)
Some college	1.32	(0.16)***	1.01	(0.17)***	1.28	(0.16)***	0.91	(0.18)***	1.14	(0.20)***	0.74	(0.22)**
College graduate or higher	1.38	(0.15)***	1.09	(0.17)***	1.78	(0.18)***	1.26	(0.20)***	1.94	(0.22)***	1.47	(0.25)***
Family structure (other=excluded)												
Two-parent	0.26	(0.10)**	0.05	(0.11)	0.03	(0.11)	-0.07	(0.12)	0.03	(0.15)	-0.02	(0.17)

(continued)

Table D6.4b (*continued*)

Variable	1972 cohort (NLS-72)				1982 cohort (HS&B)				1992 cohort (NELS)			
	Simple model		Full model		Simple model		Full model		Simple model		Full model	
	Coef.	Robust SE	Coef.	Robust SE	Coef.	Robust SE	Coef.	Robust SE	Coef.	Robust SE	Coef.	Robust SE
Math score (Quartile 1 = excluded)												
Quartile 2			1.11	(0.12)***			0.95	(0.15)***			0.76	(0.16)***
Quartile 3			2.11	(0.13)***			1.80	(0.15)***			1.29	(0.27)***
Quartile 4			3.31	(0.18)***			3.15	(0.18)***			2.87	(0.32)***
Premium (Quartile 1 = excluded)												
Quartile 2			-0.32	(0.12)**			-0.30	(0.13)*			-0.11	(0.17)
Quartile 3			-0.67	(0.13)***			-0.41	(0.14)**			-0.39	(0.21)
Quartile 4			-0.84	(0.14)***			-0.90	(0.15)***			-0.19	(0.18)
Constant	-1.29	(0.12)***	-1.96	(0.16)***	-1.33	(0.14)***	-2.08	(0.19)***	-0.30	(0.22)	-0.77	(0.32)*
Model χ^2	289.3		624.8		285.5		521.5		270.8		353.0	
Degrees of freedom	10		16		10		16		10		16	
N	3,429		3,429		3,610		3,610		4,060		4,060	

Sources: NLS 1972; HS&B 1980; NELS 1988 (see Appendix B).

Notes: Premium is the difference between the standardized math score and the standardized verbal score. Sample sizes for the HS&B and NELS cohorts are rounded to the nearest ten, in accordance with Institute of Education Statistics policy.

*p<.05. **p<.01. ***p<.001.

Table D6.5. Percentage of respondents expecting a science major given a bachelor's degree, by demographic subgroups

	Male			Female		
	1972 cohort (NLS-72)	1982 cohort (HS&B)	1992 cohort (NELS)	1972 cohort (NLS-72)	1982 cohort (HS&B)	1992 cohort (NELS)
Race						
White	35.8	41.2	26.7	14.0	12.2	9.5
Asian	43.7	47.8	38.9	14.0	25.8	11.7
Hispanic	35.3	36.6	25.6	5.2	16.3	9.2
Black	31.9	46.4	29.2	20.1	20.2	18.1
Family income						
Quartile 1	33.2	43.1	34.5	14.2	13.6	8.5
Quartile 2	39.7	42.2	29.8	13.3	12.6	10.9
Quartile 3	33.7	46.5	23.3	15.0	15.6	11.4
Quartile 4	36.2	36.1	23.4	14.3	12.8	11.0
Mother's education						
Less than high school	33.8	36.7	26.8	12.5	15.3	10.5
High school graduate	35.0	45.0	24.8	13.6	13.5	9.2
Some college	39.8	33.9	32.1	15.8	13.5	16.3
College graduate or higher	35.7	41.4	29.2	16.2	12.5	10.1
Family structure						
Two-parent	36.3	42.1	29.8	14.2	13.3	11.1
Other	33.1	37.9	21.8	14.6	14.1	9.3

(continued)

Table D6.5 (continued)

	Male			Female		
	1972 cohort (NLS-72)	1982 cohort (HS&B)	1992 cohort (NELS)	1972 cohort (NLS-72)	1982 cohort (HS&B)	1992 cohort (NELS)
Math score						
Quartile 1	17.5	20.8	20.8	4.4	5.6	6.6
Quartile 2	19.4	25.5	17.1	9.9	8.8	6.1
Quartile 3	35.6	27.3	22.6	13.8	10.9	8.3
Quartile 4	47.5	53.0	38.4	24.6	20.7	17.5
Premium						
Quartile 1	19.2	24.6	16.6	9.6	6.6	6.4
Quartile 2	37.9	41.7	31.5	15.8	10.7	12.9
Quartile 3	39.2	49.5	27.3	17.8	18.7	10.7
Quartile 4	38.1	44.0	30.6	16.0	20.6	12.9
Sample size	1,679	1,470	2,310	1,704	1,650	2,770

Sources: NLS 1972; HS&B 1980; NELS 1988 (see Appendix B).

Notes: Premium is the difference between the standardized math score and the standardized verbal score. Sample sizes for the HS&B and NELS cohorts are rounded to the nearest ten, in accordance with Institute of Education Statistics policy.

Table D6.6a. Logistic regression results predicting the expectation of science major given a bachelor's degree, for males

	1972 cohort (NLS-72)				1982 cohort (HS&B)				1992 cohort (NELS)			
	Simple model		Full model		Simple model		Full model		Simple model		Full model	
Variable	Coef.	Robust SE	Coef.	Robust SE	Coef.	Robust SE	Coef.	Robust SE	Coef.	Robust SE	Coef.	Robust SE
Race (white=excluded)												
Asian	0.34	(0.46)	0.34	(0.46)	0.31	(0.32)	0.18	(0.33)	0.35	(0.31)	0.29	(0.29)
Hispanic	0.12	(0.38)	0.63	(0.40)	-0.21	(0.23)	0.13	(0.22)	-0.22	(0.26)	0.03	(0.27)
Black	-0.07	(0.32)	0.46	(0.35)	0.24	(0.27)	0.77	(0.28)**	0.04	(0.34)	0.60	(0.35)
Family income (Quartile 1=excluded)												
Quartile 2	0.25	(0.18)	0.28	(0.19)	-0.09	(0.21)	-0.05	(0.22)	-0.41	(0.24)	-0.45	(0.25)
Quartile 3	-0.04	(0.19)	-0.02	(0.19)	0.09	(0.23)	0.10	(0.24)	-0.89	(0.26)**	-0.89	(0.27)**
Quartile 4	0.05	(0.18)	-0.01	(0.18)	-0.32	(0.22)	-0.32	(0.23)	-1.01	(0.29)**	-1.23	(0.30)***
Mother's education (less than high school=excluded)												
High school graduate	0.06	(0.19)	0.02	(0.18)	0.30	(0.28)	0.28	(0.30)	0.13	(0.30)	0.07	(0.33)
Some college	0.27	(0.23)	0.15	(0.23)	-0.10	(0.32)	-0.22	(0.34)	0.55	(0.36)	0.37	(0.39)
College graduate or higher	0.11	(0.22)	-0.02	(0.23)	0.21	(0.30)	-0.01	(0.33)	0.55	(0.32)	0.29	(0.34)
Family structure (other=excluded)												
Two-parent	0.12	(0.16)	0.11	(0.17)	0.20	(0.20)	0.19	(0.21)	0.60	(0.20)**	0.57	(0.21)**

(continued)

Table D6.6a *(continued)*

Variable	1972 cohort (NLS-72) Simple model Coef.	Robust SE	Full model Coef.	Robust SE	1982 cohort (HS&B) Simple model Coef.	Robust SE	Full model Coef.	Robust SE	1992 cohort (NELS) Simple model Coef.	Robust SE	Full model Coef.	Robust SE
Math score (Quartile 1 = excluded)												
Quartile 2			0.17	(0.29)			0.49	(0.51)			-0.03	(0.39)
Quartile 3			0.97	(0.27)***			0.60	(0.49)			0.38	(0.34)
Quartile 4			1.43	(0.27)***			1.67	(0.49)**			1.17	(0.33)***
Premium (Quartile 1 = excluded)												
Quartile 2			0.49	(0.22)*			0.33	(0.24)			0.45	(0.31)
Quartile 3			0.47	(0.22)*			0.47	(0.25)			0.28	(0.28)
Quartile 4			0.36	(0.22)			0.27	(0.25)			0.42	(0.26)
Constant	-0.85	(0.23)***	-2.14	(0.35)***	-0.60	(0.30)*	-2.07	(0.56)***	-1.17	(0.31)***	-1.98	(0.53)***
Model χ^2	6.7		94.9		13.6		82.8		26.2		85.9	
Degrees of freedom	10		16		10		16		10		16	
N	1,679		1,679		1,470		1,470		2,310		2,310	

Sources: NLS 1972; HS&B 1980; NELS 1988 (see Appendix B).

Notes: Premium is the difference between the standardized math score and the standardized verbal score. Sample sizes for the HS&B and NELS cohorts are rounded to the nearest ten, in accordance with Institute of Education Statistics policy.

$*p<.05.$ $**p<.01.$ $***p<.001.$

Table D6.6b. Logistic regression results predicting expectation of science major given a bachelor's degree, for females

| | 1972 cohort (NLS-72) | | | | 1982 cohort (HS&B) | | | | 1992 cohort (NELS) | | | |
| | Simple model | | Full model | | Simple model | | Full model | | Simple model | | Full model | |
Variable	Coef.	Robust SE	Coef.	Robust SE	Coef.	Robust SE	Coef.	Robust SE	Coef.	Robust SE	Coef.	Robust SE
Race (white=excluded)												
Asian	0.04	(0.59)	0.00	(0.62)	0.94	(0.38)*	0.80	(0.41)	0.20	(0.27)	0.19	(0.26)
Hispanic	-0.99	(0.90)	-0.46	(0.93)	0.38	(0.26)	0.80	(0.28)**	0.05	(0.35)	0.30	(0.37)
Black	0.49	(0.28)	1.24	(0.33)***	0.63	(0.28)*	1.21	(0.28)***	0.98	(0.28)***	1.32	(0.26)***
Family income (Quartile 1=excluded)												
Quartile 2	-0.08	(0.25)	-0.16	(0.25)	0.07	(0.27)	0.03	(0.28)	0.42	(0.26)	0.41	(0.25)
Quartile 3	0.05	(0.24)	-0.01	(0.25)	0.34	(0.30)	0.41	(0.31)	0.49	(0.26)	0.43	(0.27)
Quartile 4	-0.06	(0.25)	-0.15	(0.25)	0.14	(0.31)	0.09	(0.33)	0.45	(0.32)	0.29	(0.32)
Mother's education (less than high school=excluded)												
High school graduate	0.12	(0.24)	0.08	(0.25)	-0.05	(0.31)	-0.17	(0.33)	-0.28	(0.30)	-0.43	(0.33)
Some college	0.28	(0.29)	0.27	(0.30)	-0.04	(0.35)	-0.17	(0.37)	0.36	(0.35)	0.07	(0.37)
College graduate or higher	0.35	(0.29)	0.25	(0.29)	-0.13	(0.36)	-0.40	(0.38)	-0.21	(0.33)	-0.59	(0.37)
Family structure (other=excluded)												
Two-parent	0.01	(0.21)	-0.06	(0.21)	-0.02	(0.24)	-0.14	(0.25)	0.24	(0.19)	0.22	(0.19)

(continued)

Table D6.6b (*continued*)

Variable	1972 cohort (NLS-72) Simple model Coef.	Robust SE	Full model Coef.	Robust SE	1982 cohort (HS&B) Simple model Coef.	Robust SE	Full model Coef.	Robust SE	1992 cohort (NELS) Simple model Coef.	Robust SE	Full model Coef.	Robust SE
Math score (Quartile 1 = excluded)												
Quartile 2			1.15	(0.38)**			0.65	(0.49)			−0.02	(0.35)
Quartile 3			0.01	(0.26)			0.63	(0.31)*			0.10	(0.25)
Quartile 4			−0.13	(0.27)			0.70	(0.31)*			0.19	(0.26)
Constant	−1.99	(0.30)***	−3.49	(0.42)***	−2.03	(0.34)***	−3.54	(0.56)***	−2.69	(0.32)***	−3.35	(0.41)***
Model χ^2	8.1		71.2		12.1		86.6		21.0		70.1	
Degrees of freedom	10		16		10		16		10		16	
N	1,704		1,704		1,650		1,650		2,770		2,770	

Sources: NLS 1972; HS&B 1980; NELS 1988 (see Appendix B).

Notes: Premium is the difference between the standardized math score and the standardized verbal score. Sample sizes for the HS&B and NELS cohorts are rounded to the nearest ten, in accordance with Institute of Education Statistics policy.

*$p<.05$. **$p<.01$. ***$p<.001$.

Table D7.1. Descriptive statistics, by gender and cohort (percent)

	Male			Female		
	1972 cohort (NLS-72)	1982 cohort (HS&B)	1992 cohort (NELS)	1972 cohort (NLS-72)	1982 cohort (HS&B)	1992 cohort (NELS)
Race						
White	89.9	75.2	74.6	86.5	75.7	72.3
Asian	0.9	1.1	4.2	1.0	1.0	3.7
Hispanic	2.9	13.4	9.1	3.1	11.5	10.8
Black	6.3	10.3	12.1	9.4	11.9	13.2
Mother's education						
Less than high school	20.5	15.2	13.0	26.1	20.2	17.0
High school graduate	55.9	55.7	49.9	53.2	52.2	51.0
Some college	10.6	13.0	12.8	9.4	14.3	11.1
College graduate or higher	13.0	16.1	24.4	11.2	13.3	20.9
Family structure						
Two-parent	77.3	77.1	64.5	75.2	73.8	61.4
Other	22.7	22.7	35.5	24.8	26.2	38.6
Sample size	3,769	3,360	4,120	4,300	3,790	4,560

Sources: NLS 1972; HS&B 1980; NELS 1988 (see Appendix B).
Notes: Sample sizes for the HS&B and NELS cohorts are rounded to the nearest ten, in accordance with Institute of Education Statistics policy.

Table D7.2. Percentage of respondents who attained a bachelor's degree, by demographic subgroups

	Male			Female		
	1972 cohort (NLS-72)	1982 cohort (HS&B)	1992 cohort (NELS)	1972 cohort (NLS-72)	1982 cohort (HS&B)	1992 cohort (NELS)
Race						
White	29.0	35.2	34.6	25.0	34.5	41.1
Asian	43.7	49.8	47.8	54.0	52.3	48.5
Hispanic	12.0	13.9	16.7	7.8	14.4	17.0
Black	15.4	18.2	10.0	16.2	12.6	26.5
Family income						
Quartile 1	15.6	22.0	12.5	14.3	16.2	18.3
Quartile 2	23.2	25.2	25.2	19.2	27.3	32.2
Quartile 3	29.3	32.4	39.9	26.4	34.8	52.2
Quartile 4	45.2	45.9	65.4	43.1	51.5	76.2
Mother's education						
Less than high school	13.4	12.9	8.7	11.7	11.7	11.3
High school graduate	25.4	26.9	23.0	22.1	25.4	30.5
Some college	41.4	39.5	32.6	38.6	41.1	50.9
College graduate or higher	50.0	53.9	56.6	48.4	62.4	65.7

Family structure						
Two-parent	30.8	32.6	37.8	26.7	32.6	44.8
Other	17.7	24.6	17.4	15.5	21.7	24.2
Math score						
Quartile 1	4.3	5.9	5.0	3.9	6.1	5.8
Quartile 2	14.2	12.3	16.9	17.0	13.7	26.3
Quartile 3	33.7	31.3	34.9	34.5	32.9	45.2
Quartile 4	54.5	61.2	64.3	53.5	70.4	75.9
Premium						
Quartile 1	23.4	23.4	19.9	22.8	27.3	28.3
Quartile 2	26.4	31.6	28.4	25.5	33.4	37.4
Quartile 3	28.7	36.5	31.8	23.7	32.6	42.3
Quartile 4	30.3	31.3	38.4	23.8	25.8	45.0
Sample size	3,769	3,360	4,120	4,300	3,790	4,560

Sources: NLS 1972; HS&B 1980; NELS 1988 (see Appendix B).

Notes: Premium is the difference between the standardized math score and the standardized verbal score. Sample sizes for the HS&B and NELS cohorts are rounded to the nearest ten, in accordance with Institute of Education Statistics policy.

Table D7.3a. Logistic regression results predicting attainment of a bachelor's degree, for males

	1972 cohort (NLS-72)				1982 cohort (HS&B)				1992 cohort (NELS)			
	Simple model		Full model		Simple model		Full model		Simple model		Full model	
Variable	Coef.	Robust SE	Coef.	Robust SE	Coef.	Robust SE	Coef.	Robust SE	Coef.	Robust SE	Coef.	Robust SE
Race (white = excluded)												
Asian	0.65	(0.35)	0.42	(0.38)	0.67	(0.25)**	0.52	(0.27)	0.42	(0.31)	0.32	(0.33)
Hispanic	−0.50	(0.27)	0.26	(0.28)	−0.97	(0.16)***	−0.44	(0.18)*	−0.37	(0.20)	−0.08	(0.21)
Black	−0.19	(0.19)	0.88	(0.21)***	−0.65	(0.19)**	0.07	(0.21)	−1.27	(0.27)***	−0.42	(0.27)
Family income (Quartile 1 = excluded)												
Quartile 2	0.27	(0.12)*	0.18	(0.13)	−0.02	(0.16)	−0.06	(0.17)	0.37	(0.14)*	0.27	(0.16)
Quartile 3	0.40	(0.12)**	0.30	(0.13)*	0.20	(0.13)	0.17	(0.14)	0.78	(0.17)***	0.79	(0.18)***
Quartile 4	0.94	(0.12)***	0.70	(0.13)***	0.54	(0.15)***	0.38	(0.16)*	1.49	(0.20)***	1.20	(0.22)***
Mother's education (less than high school = excluded)												
High school graduate	0.55	(0.12)***	0.41	(0.12)**	0.74	(0.18)***	0.68	(0.19)***	0.74	(0.24)**	0.46	(0.26)
Some college	1.11	(0.15)***	0.79	(0.16)***	1.21	(0.21)***	0.99	(0.23)***	1.19	(0.30)***	0.76	(0.34)*
College graduate or higher	1.39	(0.15)***	1.07	(0.16)***	1.74	(0.20)***	1.30	(0.21)***	1.89	(0.25)***	1.35	(0.27)***

	Model 1	Model 2	Model 3	Model 4	Model 5	Model 6
Family structure (other=excluded)						
Two-parent	0.50 (0.10)***	0.47 (0.11)***	0.21 (0.13)	0.18 (0.15)	0.72 (0.14)***	0.61 (0.13)***
Math score (Quartile 1=excluded)						
Quartile 2		1.13 (0.17)***		0.27 (0.21)		1.11 (0.27)***
Quartile 3		2.23 (0.16)***		1.39 (0.18)***		1.89 (0.25)***
Quartile 4		3.08 (0.17)***		2.60 (0.18)***		2.95 (0.26)***
Premium (Quartile 1=excluded)						
Quartile 2		-0.43 (0.14)**		-0.60 (0.18)**		-0.26 (0.20)
Quartile 3		-0.32 (0.14)*		-0.54 (0.17)**		-0.06 (0.20)
Quartile 4		-0.59 (0.14)***		-0.84 (0.18)***		-0.06 (0.20)
Constant	-2.43 (0.14)***	-3.76 (0.22)***	-1.91 (0.20)***	-2.61 (0.24)***	-2.84 (0.30)***	-4.10 (0.36)***
Model χ^2	331.6	716.5	212.2	439.7	306.9	478.0
Degrees of freedom	10	16	10	16	10	16
N	3,769	3,769	3,360	3,360	4,120	4,120

Sources: NLS 1972; HS&B 1980; NELS 1988 (see Appendix B).

Notes: Premium is the difference between the standardized math score and the standardized verbal score. Sample sizes for the HS&B and NELS cohorts are rounded to the nearest ten, in accordance with Institute of Education Statistics policy.

*$p<.05$. **$p<.01$. ***$p<.001$.

Table D7.3b. Logistic regression results predicting attainment of a bachelor's degree, for females

| | 1972 cohort (NLS-72) | | | | 1982 cohort (HS&B) | | | | 1992 cohort (NELS) | | | |
| | Simple model | | Full model | | Simple model | | Full model | | Simple model | | Full model | |
Variable	Coef.	Robust SE	Coef.	Robust SE	Coef.	Robust SE	Coef.	Robust SE	Coef.	Robust SE	Coef.	Robust SE
Race (white = excluded)												
Asian	1.42	(0.39)***	1.44	(0.54)**	0.68	(0.28)*	0.69	(0.29)*	0.03	(0.31)	0.19	(0.30)
Hispanic	−0.82	(0.27)**	−0.07	(0.29)	−0.76	(0.15)***	−0.28	(0.17)	−0.58	(0.20)**	−0.21	(0.22)
Black	−0.08	(0.14)	0.93	(0.16)***	−1.01	(0.17)***	−0.28	(0.18)	−0.08	(0.25)	0.26	(0.33)
Family income (Quartile 1 = excluded)												
Quartile 2	0.15	(0.11)	0.05	(0.12)	0.34	(0.15)*	0.38	(0.16)*	0.26	(0.15)	0.11	(0.17)
Quartile 3	0.28	(0.11)*	0.21	(0.12)	0.57	(0.12)***	0.50	(0.14)***	0.77	(0.16)***	0.54	(0.17)**
Quartile 4	0.87	(0.12)***	0.67	(0.13)***	1.01	(0.14)***	0.84	(0.16)***	1.64	(0.19)***	1.33	(0.21)***
Mother's education (less than high school = excluded)												
High school graduate	0.58	(0.11)***	0.47	(0.11)***	0.66	(0.15)***	0.41	(0.15)**	0.88	(0.16)***	0.51	(0.18)**
Some college	1.28	(0.14)***	0.99	(0.16)***	1.33	(0.18)***	1.00	(0.19)***	1.58	(0.21)***	1.08	(0.21)***
College graduate or higher	1.53	(0.14)***	1.24	(0.15)***	2.06	(0.18)***	1.57	(0.19)***	1.93	(0.19)***	1.25	(0.21)***

	Model 1		Model 2		Model 3		Model 4		Model 5		Model 6	
Family structure (other=excluded)												
Two-parent	0.49	(0.10)***	0.35	(0.11)**	0.25	(0.12)*	0.19	(0.13)	0.57	(0.13)***	0.38	(0.14)**
Math score (Quartile 1=excluded)												
Quartile 2			1.24	(0.14)***			−0.03	(0.16)			1.54	(0.19)***
Quartile 3			2.20	(0.14)***			1.08	(0.14)***			2.19	(0.19)***
Quartile 4			3.01	(0.15)***			2.53	(0.16)***			3.41	(0.20)***
Premium (Quartile 1=excluded)												
Quartile 2			−0.32	(0.11)**			−0.38	(0.14)**			−0.20	(0.15)
Quartile 3			−0.47	(0.12)***			−0.48	(0.15)**			−0.13	(0.20)
Quartile 4			−0.72	(0.13)***			−1.00	(0.17)***			−0.26	(0.16)
Constant	−2.52	(0.13)***	−3.66	(0.18)***	−2.22	(0.17)***	−2.44	(0.20)***	−2.40	(0.18)***	−3.62	(0.26)***
Model χ^2	390.5		715.1		339.1		542.3		374.7		617.0	
Degrees of freedom	10		16		10		16		10		16	
N	4,300		4,300		3,790		3,790		4,560		4,560	

Sources: NLS 1972; HS&B 1980; NELS 1988 (see Appendix B).

Notes: Premium is the difference between the standardized math score and the standardized verbal score. Sample sizes for the HS&B and NELS cohorts are rounded to the nearest ten, in accordance with Institute of Education Statistics policy.

*$p < .05$. **$p < .01$. ***$p < .001$.

Table D7.4. Percentage of respondents majoring in science given a bachelor's degree, by demographic subgroups

	Male			Female		
	1972 cohort (NLS-72)	1982 cohort (HS&B)	1992 cohort (NELS)	1972 cohort (NLS-72)	1982 cohort (HS&B)	1992 cohort (NELS)
Race						
White	29.1	31.1	26.8	10.2	13.1	12.2
Asian	28.8	58.6	44.6	8.5	31.8	25.4
Hispanic	22.0	35.0	31.3	12.4	11.6	10.5
Black	20.1	23.7	29.1	11.3	18.6	16.7
Family income						
Quartile 1	26.5	30.8	32.4	8.2	15.5	11.9
Quartile 2	30.8	36.3	29.8	10.4	12.6	12.7
Quartile 3	31.5	36.3	26.9	8.7	13.9	13.4
Quartile 4	26.9	24.3	26.5	12.5	12.8	14.5
Mother's education						
Less than high school	28.2	28.2	29.3	4.3	15.4	14.2
High school graduate	30.9	33.3	24.9	9.7	13.2	13.4
Some college	29.0	21.6	27.0	8.8	12.8	9.8
College graduate or higher	23.7	34.8	31.4	15.7	14.6	14.3

Family structure						
Two-parent	29.6	33.0	28.4	10.5	13.7	13.7
Other	23.0	23.4	27.8	8.6	13.4	11.9
Math score						
Quartile 1	13.0	27.8	11.9	3.5	6.6	3.5
Quartile 2	14.6	8.0	13.5	5.8	11.7	7.7
Quartile 3	23.2	13.4	17.3	8.2	7.0	9.1
Quartile 4	36.9	41.5	38.8	15.7	20.9	19.3
Premium						
Quartile 1	18.2	14.1	13.1	7.8	5.2	7.8
Quartile 2	29.5	31.6	32.3	13.7	14.2	15.1
Quartile 3	29.1	38.8	32.5	10.8	20.7	15.5
Quartile 4	31.6	33.4	28.5	8.1	15.7	14.4
Sample size	1,368	1,020	1,430	1,359	1,190	1,800

Sources: NLS 1972; HS&B 1980; NELS 1988 (see Appendix B).

Notes: Premium is the difference between the standardized math score and the standardized verbal score. Sample sizes for the HS&B and NELS cohorts are rounded to the nearest ten, in accordance with Institute of Education Statistics policy.

Table D7.5a. Logistic regression results predicting science major given a bachelor's degree, for males

| | 1972 cohort (NLS-72) | | | | 1982 cohort (HS&B) | | | | 1992 cohort (NELS) | | | |
| | Simple model | | Full model | | Simple model | | Full model | | Simple model | | Full model | |
Variable	Coef.	Robust SE	Coef.	Robust SE	Coef.	Robust SE	Coef.	Robust SE	Coef.	Robust SE	Coef.	Robust SE
Race (white=excluded)												
Asian	-0.04	(0.48)	-0.18	(0.47)	1.21	(0.40)**	1.09	(0.37)**	0.75	(0.34)*	0.76	(0.31)*
Hispanic	-0.43	(0.55)	-0.01	(0.64)	0.18	(0.32)	0.50	(0.36)	0.17	(0.34)	0.37	(0.35)
Black	-0.43	(0.36)	0.10	(0.41)	-0.34	(0.39)	-0.02	(0.39)	0.09	(0.37)	0.76	(0.40)
Family income (Quartile 1=excluded)												
Quartile 2	0.11	(0.22)	0.14	(0.22)	0.15	(0.32)	0.13	(0.32)	-0.12	(0.25)	-0.18	(0.26)
Quartile 3	0.15	(0.22)	0.15	(0.22)	0.08	(0.27)	0.12	(0.29)	-0.32	(0.28)	-0.23	(0.29)
Quartile 4	-0.01	(0.21)	-0.03	(0.22)	-0.50	(0.29)	-0.58	(0.30)	-0.41	(0.30)	-0.59	(0.32)
Mother's education (less than high school=excluded)												
High school graduate	0.04	(0.23)	0.07	(0.24)	0.24	(0.42)	0.28	(0.48)	-0.03	(0.42)	-0.33	(0.39)
Some college	-0.05	(0.28)	-0.10	(0.28)	-0.23	(0.47)	-0.34	(0.53)	0.11	(0.46)	-0.26	(0.44)
College graduate or higher	-0.29	(0.28)	-0.37	(0.28)	0.42	(0.45)	0.36	(0.51)	0.38	(0.43)	-0.08	(0.40)

	(1)		(2)		(3)		(4)		(5)		(6)	
Family structure (other = excluded)												
Two-parent	0.28	(0.20)	0.20	(0.21)	0.49	(0.27)	0.39	(0.29)	0.08	(0.23)	0.09	(0.24)
Math score (Quartile 1 = excluded)												
Quartile 2			0.36	(0.48)			-1.47	(0.66)*			0.26	(0.52)
Quartile 3			0.93	(0.46)*			-0.87	(0.47)			0.55	(0.48)
Quartile 4			1.60	(0.46)***			0.61	(0.39)			1.66	(0.48)**
Premium (Quartile 1 = excluded)												
Quartile 2			0.22	(0.27)			0.69	(0.37)			0.49	(0.34)
Quartile 3			0.17	(0.26)			0.90	(0.38)*			0.56	(0.32)
Quartile 4			0.23	(0.26)			0.86	(0.39)*			0.37	(0.31)
Constant	-1.14	(0.29)***	-2.48	(0.53)***	-1.32	(0.46)**	-2.18	(0.62)***	-0.99	(0.44)*	-2.22	(0.66)**
Model χ^2	10.9		63.2		28.0		77.2		10.8		70.2	
Degrees of freedom	10		16		10		16		10		16	
N	1,368		1,368		1,020		1,020		1,430		1,430	

Sources: NLS 1972; HS&B 1980; NELS 1988 (see Appendix B).

Notes: Premium is the difference between the standardized math score and the standardized verbal score. Sample sizes for the HS&B and NELS cohorts are rounded to the nearest ten, in accordance with Institute of Education Statistics policy.

*$p < .05$. **$p < .01$. ***$p < .001$.

Table D7.5b. Logistic regression results predicting science major given a bachelor's degree, for females

	1972 cohort (NLS-72)				1982 cohort (HS&B)				1992 cohort (NELS)			
	Simple model		Full model		Simple model		Full model		Simple model		Full model	
	Coef.	Robust SE	Coef.	Robust SE	Coef.	Robust SE	Coef.	Robust SE	Coef.	Robust SE	Coef.	Robust SE
Race (white=excluded)												
Asian	−0.08	(0.77)	−0.05	(0.78)	1.09	(0.39)**	1.08	(0.43)*	0.87	(0.29)**	0.99	(0.27)***
Hispanic	0.51	(0.86)	0.97	(0.91)	−0.16	(0.34)	0.09	(0.35)	−0.15	(0.36)	−0.02	(0.42)
Black	0.49	(0.35)	1.16	(0.41)**	0.39	(0.39)	0.82	(0.42)*	0.49	(0.46)	0.82	(0.42)*
Family income (Quartile 1=excluded)												
Quartile 2	0.23	(0.31)	0.18	(0.32)	−0.23	(0.36)	−0.33	(0.37)	0.14	(0.37)	0.14	(0.37)
Quartile 3	−0.12	(0.33)	−0.13	(0.33)	−0.15	(0.32)	−0.22	(0.34)	0.18	(0.33)	0.16	(0.34)
Quartile 4	0.15	(0.32)	0.06	(0.33)	−0.26	(0.36)	−0.43	(0.37)	0.30	(0.35)	0.22	(0.35)
Mother's education (less than high school=excluded)												
High school graduate	0.99	(0.39)*	0.95	(0.39)*	−0.06	(0.42)	−0.13	(0.43)	−0.09	(0.36)	−0.18	(0.39)
Some college	0.92	(0.45)*	0.91	(0.46)*	−0.06	(0.46)	−0.10	(0.48)	−0.48	(0.42)	−0.70	(0.46)
College graduate or higher	1.52	(0.45)**	1.42	(0.45)**	0.07	(0.45)	−0.11	(0.46)	−0.10	(0.39)	−0.35	(0.43)

	Model 1		Model 2		Model 3		Model 4		Model 5		Model 6	
Family structure (other=excluded)												
Two-parent	0.23	(0.31)	0.21	(0.32)	0.11	(0.30)	0.11	(0.31)	0.15	(0.26)	0.03	(0.26)
Math score (Quartile 1=excluded)												
Quartile 2			0.68	(0.56)			0.72	(0.75)			0.76	(0.86)
Quartile 3			1.10	(0.53)*			0.29	(0.63)			1.09	(0.80)
Quartile 4			1.82	(0.54)**			1.52	(0.64)*			1.96	(0.79)*
Premium (Quartile 1=excluded)												
Quartile 2			0.20	(0.27)			0.41	(0.52)			0.23	(0.29)
Quartile 3			-0.09	(0.30)			0.80	(0.53)			0.31	(0.35)
Quartile 4			-0.37	(0.34)			0.42	(0.53)			0.16	(0.30)
Constant	-3.51	(0.49)***	-4.75	(0.70)***	-1.80	(0.49)***	-3.15	(0.61)***	-2.12	(0.43)***	-3.56	(0.96)***
Model χ^2	21.2		49.7		11.3		57.6		14.6		62.4	
Degrees of freedom	10		10		10		16		10		16	
N	1,359		1,359		1,190		1,190		1,800		1,800	

Sources: NLS 1972; HS&B 1980; NELS 1988 (see Appendix B).

Notes: Premium is the difference between the standardized math score and the standardized verbal score. Sample sizes for the HS&B and NELS cohorts are rounded to the nearest ten, in accordance with Institute of Education Statistics policy.

* $p<.05$. ** $p<.01$. *** $p<.001$.

Table D7.6. Transition rates from science degree to further education, by gender, field, and cohort (percent)

Level of degree attained in previous 2 years	NES						NSRCG	
	1976	1980	1982	1984	1986	1988	2003	2006
Bachelor's degree								
Both sexes	44.8	36.9	33.7	35.5	29.6	30.0	32.0	29.6
Biological	48.2	42.4	41.2	46.1	44.5	42.6	43.7	43.9
Engineering	34.4	26.9	24.3	28.8	23.5	24.2	24.8	23.8
Physical	61.0	48.2	49.3	47.9	44.0	42.2	46.5	42.8
Mathematical	42.6	30.0	24.0	24.0	18.9	21.8	18.8	15.5
Male	44.8	38.5	32.9	36.1	30.6	29.9	29.8	26.0
Biological	49.4	46.8	44.0	52.0	52.4	45.2	45.3	45.9
Engineering	34.5	26.7	23.8	28.4	23.3	24.1	24.5	23.0
Physical	61.6	51.6	50.6	50.4	45.2	43.6	47.8	43.5
Mathematical	43.0	36.2	23.2	26.4	21.0	22.2	18.7	13.7
Female	44.8	32.3	35.6	33.9	27.3	30.3	35.4	35.3
Biological	44.8	35.0	37.7	39.7	35.2	39.5	42.6	42.6
Engineering	30.3	28.9	28.9	31.5	25.1	24.8	26.0	26.4
Physical	57.6	38.1	46.0	41.8	40.9	38.0	45.1	41.9
Mathematical	41.8	18.8	25.6	20.4	15.7	21.2	18.9	20.3

Master's degree

Both sexes								
Biological	45.3	39.0	41.0	41.9	37.9	28.3	37.8	28.6
Engineering	22.7	22.9	24.2	25.6	24.6	27.5	26.5	27.1
Physical	37.3	35.4	41.5	39.8	41.3	39.6	39.1	47.5
Mathematical	32.0	21.1	26.7	19.2	16.4	18.3	24.1	22.5
	31.4	28.5	31.4	29.8	27.1	26.6	28.8	27.3
Male								
Biological	46.1	41.1	43.8	46.6	42.1	28.7	45.3	32.0
Engineering	21.9	22.0	24.6	26.4	24.9	28.6	26.7	27.3
Physical	42.1	39.5	44.1	43.7	43.1	43.3	48.9	53.6
Mathematical	*34.0*	24.5	27.6	18.7	18.9	21.8	24.4	25.1
	31.4	29.0	31.9	30.9	28.0	28.8	30.0	29.0
Female								
Biological	*42.9*	34.4	36.3	34.1	32.9	27.8	30.9	25.8
Engineering	*41.1*	36.0	21.0	19.0	22.1	19.2	25.8	26.1
Physical	*10.0*	19.2	32.0	27.7	35.3	28.1	23.9	36.0
Mathematical	*24.6*	13.3	24.3	20.3	9.3	9.5	23.6	17.3
	31.6	26.6	29.7	26.0	23.9	19.4	26.2	23.8

Sources: NES 1976–1988; NSRCG 2003, 2006 (see Appendix C).
Notes: Entries are transition rates for inflow into (State 1 + State 2 + State 3). The states are defined in Figure 7.3. Estimates based on fewer than 100 cases are presented in italics.

Table D7.7. Transition rates from science degrees to specific education programs, by gender and field, 2003–2006

Level of degree attained in previous 2 years	2003				2006			
	All (states 1–3)	S/E (state 1)	Professional (state 2)	Others (state 3)	All (states 1–3)	S/E (state 1)	Professional (state 2)	Others (state 3)
Bachelor's degree								
Both sexes	32.0	15.5	8.7	7.8	29.6	13.4	9.4	6.7
Biological	43.7	11.5	19.4	12.8	43.9	12.2	22.4	9.3
Engineering	24.8	18.9	1.5	4.4	23.8	17.5	2.8	3.5
Physical	46.5	29.9	10.7	5.9	42.8	27.0	8.1	7.7
Mathematical	18.8	13.0	0.7	5.1	15.5	7.8	1.3	6.4
Male	29.8	16.6	7.3	5.9	26.0	13.4	7.6	4.9
Biological	45.3	11.4	23.7	10.2	45.9	12.0	27.5	6.4
Engineering	24.5	19.3	1.2	4.0	23.0	17.2	2.5	3.4
Physical	47.8	34.1	9.3	4.4	43.5	29.3	7.8	6.5
Mathematical	18.7	13.1	0.6	5.0	13.7	7.6	1.0	5.1
Female	35.4	14.0	10.8	10.6	35.3	13.5	12.2	9.6
Biological	42.6	11.6	16.5	14.5	42.6	12.3	19.2	11.1
Engineering	26.0	17.3	2.8	5.8	26.4	18.5	3.8	4.1
Physical	45.1	25.2	12.2	7.7	41.9	23.9	8.6	9.3
Mathematical	18.9	12.7	0.9	5.3	20.3	8.5	2.1	9.6

Master's degree

Both sexes	28.8	20.6	2.9	5.2	27.3	22.1	1.6	3.6
Biological	37.8	19.7	12.1	6.1	28.6	19.1	6.2	3.2
Engineering	26.5	21.6	1.3	3.6	27.1	22.7	1.2	3.1
Physical	39.1	34.1	1.2	3.8	47.5	44.2	1.1	2.1
Mathematical	24.1	16.2	0.5	7.3	22.5	17.7	0.1	4.7
Male	30.0	22.2	2.8	5.1	29.0	23.6	1.5	3.9
Biological	45.3	22.3	16.7	6.4	32.0	23.4	6.5	2.1
Engineering	26.7	21.6	1.0	4.1	27.3	22.6	1.3	3.4
Physical	48.9	44.1	0.0	4.8	53.6	50.8	1.1	1.7
Mathematical	24.4	17.4	0.8	6.2	25.1	19.2	0.1	5.8
Female	26.2	17.5	3.2	5.5	23.8	19.0	2.0	2.8
Biological	30.9	17.3	7.8	5.8	25.8	15.7	6.1	4.1
Engineering	25.8	21.4	2.5	1.8	26.1	22.9	1.0	2.2
Physical	23.9	18.5	3.1	2.2	36.0	32.0	1.1	2.9
Mathematical	23.6	14.3	0.2	9.1	17.3	14.8	0.1	2.4

Sources: NSRCG 2003, 2006 (see Appendix C).
Notes: Entries are transition rates for inflow into State 1, State 2, or State 3. The states are defined in Figure 7.3.

Table D8.1. Transition rates from science degree to science occupation, conditional on working, by field, gender, and cohort (percent)

Level of degree attained in previous 2 years	NES						NSRCG	
	1976	1980	1982	1984	1986	1988	2003	2006
Bachelor's degree								
Both sexes	60.9	72.7	82.5	81.9	80.4	81.5	54.1	54.0
Biological	44.5	53.7	63.3	63.9	55.5	59.3	26.7	19.2
Engineering	81.8	92.3	93.1	92.2	90.4	92.1	78.1	79.1
Physical	55.4	64.7	77.1	64.9	65.4	65.9	46.4	44.4
Mathematical	54.7	75.7	88.3	86.8	83.9	84.4	58.0	59.0
Male	63.9	76.9	86.0	85.6	83.9	84.5	62.7	64.4
Biological	45.8	53.5	67.1	69.4	62.2	64.6	29.2	27.9
Engineering	81.8	92.1	93.2	92.3	90.8	92.4	79.4	80.6
Physical	55.2	65.3	79.1	69.1	65.6	67.5	51.7	45.5
Mathematical	56.5	78.3	89.5	87.5	85.2	84.9	64.0	65.3
Female	47.1	49.0	71.8	73.2	72.0	74.1	40.0	34.0
Biological	41.0	54.1	58.7	58.7	49.4	53.6	25.0	14.1
Engineering	84.3	94.1	92.0	91.6	87.8	90.4	72.9	73.3
Physical	56.3	63.3	72.2	56.1	64.9	61.4	40.7	42.8
Mathematical	51.1	72.0	85.5	85.8	82.2	83.7	45.5	39.7

Master's degree

Both sexes	82.5	86.9	89.2	90.7	90.1	88.8	80.8	76.9
Biological	*71.7*	74.8	78.7	80.5	80.7	73.4	57.2	52.2
Engineering	93.5	94.2	94.6	95.4	94.1	93.0	89.9	84.8
Physical	*60.0*	84.3	83.6	87.4	84.2	86.5	75.7	69.0
Mathematical	*76.6*	83.7	91.0	90.5	90.4	90.9	80.8	78.4
Male	83.5	83.7	90.1	92.3	91.6	91.3	85.1	81.0
Biological	*69.2*	73.4	80.3	83.0	86.7	80.0	65.2	57.0
Engineering	94.2	94.6	95.1	95.1	94.0	93.0	90.9	85.5
Physical	*57.5*	87.2	82.9	91.5	86.8	88.8	78.0	74.6
Mathematical	76.8	85.9	89.4	91.8	90.5	93.7	82.7	81.9
Female	*74.8*	60.3	85.7	85.3	84.9	80.9	71.8	68.0
Biological	*79.5*	*77.9*	76.2	76.5	73.5	*64.1*	51.0	48.4
Engineering	*68.9*	87.1	89.8	97.2	95.0	93.0	85.9	82.3
Physical	*70.0*	*75.6*	*86.1*	76.9	75.9	80.7	*73.2*	60.3
Mathematical	*75.9*	*78.9*	*95.0*	87.8	90.2	84.6	77.2	71.5

Sources: NES 1976–1988; NSRCG 2003, 2006 (see Appendix C).
Notes: Entries are transition rates calculated as (State 4)/(State 4 + State 5). The states are defined in Figure 7.3.
Estimates based on fewer than 100 cases are presented in italics.

Notes

Introduction

1. Romer (1990, p. S72); also see Diamond (1999), Neal et al. (2008), National Academy of Sciences et al. (2007, 2010), and Solow (1957).
2. Ben-David (1971); Price (1963).
3. Bell (1976) and Rooney et al. (2005).
4. Neal et al. (2008).
5. Research!America (2007).
6. Solow (1957). Although Solow's model may be considered simplistic for assuming technological change to be exogenous, more recent efforts have aimed largely at refining his approach rather than challenging the basic premise that technological change is central to economic growth. See also Mankiw (2003).
7. Kaufman (2009).
8. Economists call this type of good a "nonrival good" (Romer 1990; Warsh 2006).
9. Concern over scientists having unequal access to journal articles has led, in recent years, to the creation of online repositories such as PubMed Central in medicine, CiteSeer[x] in computer and information science, and arXiv in Physics.
10. Economists call such effects "externalities" or "spillovers" (Mankiw 2003; Pindyck and Rubinfeld 2005).
11. Goldin and Katz (2008); National Academy of Sciences et al. (2007).
12. Goldin and Katz (2008); National Academy of Sciences et al. (2007).
13. Yuasa (1962, p. 70, emphasis added).
14. The following statistics come from Galama and Hosek (2008), except as otherwise noted.
15. Our calculation is based on data from Wikipedia (2010) and Nobelprize.org (2010). Note that this number is lower than the 70 percent cited by Galama and Hosek (2008, p. xvi) and the 60 percent since the 1930s cited by Cole (2009, p. 4) because we restricted ourselves to Nobel Prizes awarded only in the three science fields of physics, chemistry, and physiology/medicine. Americans' share of winners of the Nobel Prize in Economics is much higher.
16. See Cole (2009, p. 3).
17. IMD International (2005); Adams (2007).

18. Pew Research Center (2009). The sample was 2,533 members of the American Association for the Advancement of Science, 81 percent U.S. born, 9 percent foreign-born U.S. citizens, and 9 percent noncitizens.
19. Ibid.
20. Hollingsworth et al. (2008).
21. Atkinson (1990). See also Grogan (1990). Atkinson's *Science* article has been widely criticized, and he himself later admitted that some of its assumptions were flawed, especially where immigration was concerned (Atkinson 1996).
22. Abelson (1990).
23. While concern over an inadequate scientific labor force in the United States goes back to the 1950s (e.g., Mead and Métraux 1957), the current "crisis" debate received widespread attention beginning in the 1990s. See Neal et al. (2008, pp. 277–293) for a recent review of the debate on the shortage of scientists and engineers in the United States amounting to a possible crisis.
24. National Academy of Sciences et al. (2007).
25. This is to be discussed in Chapters 2 and 6.
26. Freeman (2006) reports that in 2000, 39 percent of science and engineering PhDs went to individuals born outside of the United States. Matthews (2008), in an issue brief for the Congressional Research Service, reports that in 2005, foreign students earned 34.7 percent of doctorates in science and 63.1 percent of those in engineering. Students on temporary visas, 56 percent of whom remain in the United States, earned 30.8 percent of science and 58.6 percent of engineering doctorates. We will discuss this topic in Chapters 4 and 7.
27. National Science Foundation (1986, p. iii).
28. Committee on Underrepresented Groups and the Expansion of the Science and Engineering Workforce Pipeline et al. (2011, chapter 1).
29. Galama and Hosek (2007).
30. Galama and Hosek (2008).
31. For a critique of the RAND report, see Ezell and Atkinson (2008).
32. Lowell and Salzman (2007).
33. Butz et al. (2003). Later in the book, we will also show empirical evidence consistent with this claim.
34. Austin (2002); Garrison et al. (2005); Teitelbaum (2002).
35. Benderly (2010); National Research Council (2005).
36. Education for Innovation Initiative (2008).
37. Obama (2009).
38. Cicerone (2009).
39. National Academy of Sciences et al. (2010, p. 4). The subtitle of the report is "Rapidly Approaching Category 5."
40. National Academy of Sciences et al. (2010).
41. Kevles (1978); Neal et al. (2008).
42. This is sometimes referred to more precisely as "mature science" or "modern experimental science," in contrast with methods of studying nature in ancient cultures and in medieval universities. The word "science" also historically meant general knowledge and skill, particularly in medieval times (*Oxford English Dictionary* 2010). While our definition concurs with the commonplace understand-

ing of science, some historians of science have cautioned that there has never been a coherent notion of science even among practicing scientists. For example, Shapin (1996, p. 3) defines science merely as a "diverse array of cultural practices aimed at understanding, explaining, and controlling the natural world" without reference to methods.

43. Citro and Kalton (1989); Xie and Shauman (2003).

44. The decision to exclude social science had important consequences for the study, as shown in Chapters 7 and 8.

45. On the website for this book (found at www.yuxie.com), we provide detailed codes that we use to define scientific occupations and science education by major field, for all data sets that we use.

1. The Evolution of American Science

1. We will discuss the estimated size of the American scientific labor force in Chapter 4.

2. National Science Board (2008, p. 4-5).

3. National Science Foundation (2010) and National Institutes of Health (2010).

4. National Science Board (2008, table 3-7). We will examine scientists' earnings in depth in Chapter 4.

5. We will discuss science degree attainment in Chapter 7 and utilization of science education in Chapter 8.

6. Merton (1942).

7. Shapin (1996). This statement needs qualifying. While it was true that the "scientists" who led experimental research were all gentlemen of independent means, lab technicians, who had specialized expertise and made important contributions to experimental research, were paid for their services (Shapin 1996, 2008).

8. A good account of the history can be found in Kuhn (1992).

9. Kuhn (1996).

10. Ben-David (1971).

11. Ibid.

12. For historical accounts of this change, see Shapin (1994, 1996) and Shapin and Schaffer (1985).

13. This new approach is apparent in the stated objective of the founders of the Royal Society: "for the promoting of Physico-Mathematicall Experimentall Learning" (quoted in Hall 1991, p. 9; Wikipedia 2009). For a history of the Royal Society, see Hall (1991).

14. Ben-David (1971). Shapin (1996) points out four interrelated changes during this period: (1) the mechanization of nature (the use of mechanical metaphors to construe natural processes and phenomena), (2) the depersonalization of natural knowledge (growing separation between human subjects and natural objects), (3) an attempted mechanization of knowledge (proposed deployment of explicitly formulated rules of method), and (4) the pursuit of disinterested knowledge, which was separate from moral, social, and political agendas.

15. For good sources on Isaac Newton's contributions, see Gleick (2003), Koyré (1965), Kuhn (1992), and Shapin (1996).

16. Newton's *Philosophiæ Naturalis Principia Mathematica* (1687) described universal gravitation and the three laws of motion, providing the foundation of classical mechanics (Kuhn 1992).
17. Clark (1983, p. 61) titles his chapter on Franklin's electricity experiments "The Professional Amateur."
18. Isaacson (2003).
19. Biagioli (1993, p. 104).
20. Wikipedia (2011); see also Freeman (1982) and Gruber (1981).
21. Bulmer (2003).
22. Ben-David (1971); Hall (1991); Merton (1970); and Shapin (1996).
23. Clark (2003); Hall (1991); Merton (1970); and Shapin (1996).
24. Ben-David (1971, p. 78) and Hahn (1971).
25. Hahn (1971).
26. Ben-David (1971).
27. Clark (2003); Mertz (1965).
28. Mertz (1965).
29. For example, in Gascoigne's (1995) study of 614 scientists born between 1660 and 1760, 69 percent did not serve as professors or in a similar capacity. Another study by Roche (1978, p. 197) arrives at the same conclusion, but his study provides more information regarding the specific backgrounds of practitioners. For instance, out of six thousand members of provincial academies, Roche found that 37 percent were nobles, 20 percent higher clergy, and 43 percent were commoners.
30. For a similar definition, see Starr (1982).
31. In Ben-David's words, "Scientific achievements were considered as being sacred, as expressions of the deepest and most essential qualities of a specially gifted person. . . . The professor was not in principle paid for research, but he occupied a role with a stipend which made it possible for him to do research as he wished" (Ben-David 1971, p. 156).
32. Of course, parallel changes occurred in other countries. We focus on this change in the United States for two reasons. First, the professionalization of science in the United States was notable for its large-scale completeness, which was responsible for the success of American science (discussed later). Second, this book is primarily concerned with science in America.
33. Our calculation is based on statistics released by the U.S. Census Bureau (2008).
34. For a study of the transformation of American universities into world-class research institutions during the period, see J. Cole (2009).
35. Ben-David (1971, p. 155).
36. Hughes (2004).
37. Twain ([1889] 1996).
38. Isaacson (2003, p. 130) writes that Franklin's goal was always to make science useful, "just as Poor Richard's wife had made sure that he did something practical with all his old 'rattling traps.' In general, he would begin a scientific inquiry driven by pure intellectual curiosity and then seek a practical application for it."
39. Bellis (2010).
40. Carlson (1997). Along with the institutionalization of scientific creativity came greater scientific and industrial standardization. For example, the patent process was also reformed during this period (Noble 1977).

41. Noble (1977, pp. 257–320).
42. Braverman (1974).
43. Hughes (2004, p. 53), for example, argues that independent inventors "invented a disproportionate share of the radical inventions," a phenomenon that owes much to the freedom and choice of such inventors.
44. Bellis (2010).
45. Kevles (1978).
46. Isaacson (2007, pp. 425–433). Also see J. Cole (2009).
47. Gosling (2010); Kevles (1978).
48. Wang (2008).
49. Obama (2011).
50. Quoted in National Aeronautics and Space Administration, Historical Staff (1963, p. 2).
51. National Aeronautics and Space Administration, Historical Staff (1963).
52. Ibid., p. 9.
53. Ibid., p. 19.
54. DARPA (2011). ARPA is now DARPA, the Defense Advanced Research Projects Agency, which played a key role in the development of the global Internet.
55. Wang (2008, p. 82).
56. Quoted in National Aeronautics and Space Administration, Historical Staff (1963, p. 6).
57. Quoted in ibid., p. 15.
58. Ben-David (1971, p. 166).
59. Kevles (1977, 1978). Vannevar Bush, a science advisor to President Franklin D. Roosevelt, was instrumental in the establishment of the National Science Foundation. See Bush (1945) and J. Cole (2009).
60. The amounts are not adjusted for inflation. Data provided by National Science Foundation (2010).
61. Data provided by National Institutes of Health (2010). Congressional appropriations of NSF and NIH budgets have frequently been subject to heated political debates in recent years.
62. Kevles (1978).
63. Ibid.
64. For an account of this change in social studies of science, see S. Cole (2004).
65. Xie and Shauman (2003).
66. Harding (1991); Keller (1996).
67. In what follows, we borrow heavily from Derek Price's work (Price 1963). See a contrarian view on the importance of individual scientists in the same period in Shapin (2008).
68. Price (1963, p. 93).
69. Ibid., p. 19.
70. Ibid., pp. 109–110.
71. Ibid., p. 104.
72. Shapin (1994).
73. Here, we borrow Kuhn's (1996) classic distinction between normal science and revolutionary science.

74. Cole and Cole (1973, p. 45).
75. Price (1963).

2. American Science and Globalization

1. Yuasa (1962).
2. Galama and Hosek (2007); Hollingsworth et al. (2008).
3. One major proponent of globalization being a strong social force is Friedman (2005).
4. See Friedman (2005).
5. Again, this feature is called "nonrivalry" by economists (Romer 1990).
6. As will be discussed in Chapter 3, this is called the "universalism hypothesis" in sociology. See Merton (1942).
7. Somsen (2008) has shown how the precise meaning of universality has varied from one historic period to another; Livingstone (2003) also points out that geographic locations continually affect both scientific studies and the degree of enthusiasm with which results are received.
8. Blanchard et al. (2009); Bound et al. (2009); Freeman (2009).
9. National Science Board (2008, p. 5-2).
10. Ibid., p. 5-7.
11. Diamond (1999).
12. In his letter to Robert Hooke, February 5, 1676 (Newton 1676).
13. Shapin (1996); Kuhn (1992).
14. Page (2007).
15. For example, medical scientists in one country may have better clinical data, whereas medical scientists in another country may have better data on animals.
16. Oyer (2007).
17. National Academy of Sciences et al. (2007).
18. OECD (2010).
19. Dillon (2010); Bruce (2010).
20. Gonzales et al. (2009); OECD (2007).
21. National Science Board (2010, p. 1-16).
22. Heston et al. (2009).
23. Kuwait is an outlier because of its very high GDP and poor test results and is removed from the data set in order to avoid unduly influencing the results.
24. The lowess regression fits a flexible but smoothed line through the points of the scatterplot.
25. For brevity, we do not present the TIMSS results for eighth graders, although they tell much the same story: students in the United States perform slightly worse than expected, given their financial resources, although the difference is larger in math.
26. Kuwait and Qatar are outliers because of their very high GDP and poor test results. They are removed from the analysis so that they will not exert undue influence on the regression line.
27. Gonzales et al. (2009).
28. Fleischman et al. (2010).

29. A detailed description of the construction of the journal database is provided in the original report, which is available online from the National Science Foundation, Division of Science Resource Statistics (2006).
30. "The crude size of science in manpower or in publications tends to double within a period of 10 to 15 years" (Price 1963, p. 6). Price assumed exponential growth rates. Under exponential growths, doubling in ten and fifteen years corresponds to about 7 percent and 5 percent annual growth rates, respectively.
31. We do not know why the trend in mathematics is different. This may be due to data not being comparable over time. In the series, we note a large, sudden drop in mathematics output in 1990. Also, by the NSF definition, computer science is part of mathematics.
32. Price (1963, chapter 1).
33. The EU-15 includes Austria, Belgium, Denmark, Finland, France, Germany, Greece, Ireland, Italy, Luxembourg, Netherlands, Portugal, Spain, Sweden, and the United Kingdom.
34. Data were drawn from Hill et al. (2007).
35. Ibid.
36. Ibid. (appendix table 6).
37. Adams and Pendlebury (2010).
38. Ibid.

3. Why Do People Become Scientists?

1. The methodological tradition of path analysis and structural equations—as exemplified in Blau and Duncan (1967), Duncan et al. (1972), and Sewell and Hauser (1975)—treats occupation as a continuous variable by using Duncan Socioeconomic Index scores, thus losing sight of the particularities of occupations. For a recent exception, see Weeden and Grusky (2005).
2. J. Davis (1964, 1965); Daymont and Andrisani (1984); Jones et al. (2000); Konrad et al. (2000).
3. Irreversibility is an important element in Ginzberg and his associates' (1951) psychological explanation of occupational choice as a developmental process. Here we wish to stress the skill-training difficulty and financial cost associated with reversing one's career choice.
4. Merton (1942, pp. 270–273). The universalism hypothesis is about the normative behavior of scientists. Admittedly, deviations from the norm of universalism are not uncommon in science. But the fact that they are considered by the scientific community as "deviant" is evidence for the claim that universalism is a norm in science. For a discussion of deviant behavior in science, see Hagstrom (1965). A well-known example of deviation from universalism in science is the forced exodus of Jewish scientists from Nazi Germany, an event that strengthened the scientific enterprise in the United States during and after World War II (J. Cole 2009). Empirical evidence of the universalism hypothesis in American science is mixed. While J. Cole and Cole (1973) and J. Cole (1979) find support for the hypothesis, works by Long et al. (1979) and Allison and Long (1987) reveal evidence against the hypothesis.

5. Cole and Cole (1973).
6. Blau and Duncan (1967); Jencks et al. (1972); Featherman and Hauser (1978); Mare (1980); Brand and Xie (2010).
7. This implication in the literature on the recruitment of scientists is found in Visher (1947), Roe (1953), Xie (1989a, 1989b), and Zuckerman (1977).
8. Hout (1988); Bell (1976).
9. For reviews of the trends in earnings returns to college education, see Katz and Autor (1999); Fischer and Hout (2006). We will compare the earnings of scientists with those of other professionals between 1960 and 2000 in Chapter 4.
10. Hodge et al. (1964); Reiss (1961, appendix B); Stevens and Featherman (1981).
11. National Science Board (2008, p. 7-33). We discuss scientists' prestige in more detail in Chapter 5.
12. We will provide more precise estimates of the scientific labor force in Chapter 4.
13. Price (1963); Cole and Cole (1973). This relationship can best be summarized by the so-called Lotka's law (Price 1963, p. 43): the number of people producing n papers sharply decreases in proportion to $1/n^2$.
14. Price (1963, p. 59).
15. Merton (1968). The "Matthew effect" commonly means that "the rich get richer and the poor get poorer." The phrase came from Merton's quote of a statement attributed to Saint Matthew in the Christian Bible: "For unto every one that hath shall be given, and he shall have abundance: but from him that hath not shall be taken away even that which he hath" (Matthew 25:29).
16. This is different from a concentration of wealth, which is primarily consumed privately for individuals' own personal benefits and comforts.
17. Cole and Cole (1973); DiPrete and Eirich (2006); J. Hirsch (2005); Vinkler (2007); Zuckerman (1977).
18. Holt and Laury (2002).
19. Freeman (2006). We will discuss this possibility in Chapters 8 and 9.
20. Galton (1874).
21. For reviews, see Eiduson and Beckman (1973) and Rever (1973).
22. Berry (1981); Mumford et al. (2005); Roe (1953); Visher (1947); Zuckerman (1977).
23. W. Hirsch (1968, p. 9).
24. West (1960, 1961); Eiduson (1962).
25. For an intellectual history of Galton's statistics, see Hilts (1973, 1981) and MacKenzie (1981). For his original writings on the subject, see Galton (1869, 1874, p. 12).
26. Galton (1874).
27. Merton (1970).
28. Weber (1958).
29. Nisbett (2009).
30. See extended discussions in Flynn (1991), Nisbett (2009), and Stevenson and Stigler (1992).
31. See Xie and Goyette (2003).

4. American Scientists

1. The full occupational classification scheme for occupations is given in Codebook A, available online at http://www.yuxie.com.
2. National Academy of Sciences et al. (2007, p. 49).
3. Freeman (2006); Matthews (2008).
4. Freeman (2006); Matthews (2008).
5. In 2002, for example, the average expenditure per college student was $2,486 in India and $6,047 in South Korea, in contrast to $20,545 in the United States (OECD 2005, table 2.13, p. 203).
6. Matthews (2008); North (1995).
7. See also Xie and Shauman (2003).
8. Authors' calculation, based on 1960–2000 U.S. Census and 2006–2008 ACS data.
9. Ibid.
10. Note that the occupations included in mathematical sciences are heterogeneous. See Codebook A, available online at http://www.yuxie.com.
11. A nationwide debate about this was sparked by comments made by then Harvard president Lawrence Summers in 2005 (Lawler 2005).
12. Xie and Shauman (2003).
13. Becker (1981); Waite (1995).
14. Authors' calculation, based on 1960–2000 U.S. Census and 2006–2008 ACS data.
15. Freeman (2006) reports that African Americans and Latinos increased their percentage of doctoral degrees in science and engineering among U.S. citizens or residents from 2 percent in 1976 to 10 percent in 2001.
16. Authors' calculation, based on 1960–2000 U.S. Census and 2006–2008 ACS data.
17. Ibid.
18. Ibid. Xie and Goyette (2004, p. 10) reported a very large jump, from 19 percent to 37 percent, in the proportion of Asians with at least a college education between 1960 and 1970, mainly attributable to high educational attainment among new Asian immigrants, whose numbers increased following the passing of the 1965 Immigration Act.
19. Bean and Leach (2005); Lee and Bean (2004).
20. Authors' calculation, based on 1960–2000 U.S. Census and 2006–2008 ACS data.
21. See Spain and Bianchi (1996) for discussions of women's labor force participation.
22. Specifically, the classic discrete choice model in economics (McFadden 1974; Train 2003), rational choice theory in sociology (Coleman 1990; Xie and Shauman 1997), and social learning theory in psychology (Bandura 1986). Empirical work utilizing this framework for the study of occupational choice can be found in Boskin (1974), Freeman (1971), Manski and Wise (1983), and Xie and Shauman (1997).
23. Of course, male nurses are much fewer in number than female nurses and thus may not represent nurses in general. However, a similar trend does hold for female nurses, shown in Appendix Table D4.3.

24. Our calculation was based on 1960–2000 U.S. Census and 2006–2008 ACS data. We adjusted the earnings data over time for inflation by using the urban Consumer Price Index (Bureau of Labor Statistics 2010a).
25. We do not have information about whether advanced degrees prior to 1980 were master's degrees or doctoral degrees. In our analysis of relative wages, we make some simplifying assumptions in order to decompose the relative wages of those with master's degrees and those with doctoral degrees (see Appendix A). Here, we calculate trends in absolute earnings across years when graduate degrees are measured more precisely.
26. For example, Borjas (2005); North (1995).
27. Authors' calculation, based on 1960–2000 U.S. Census and 2006–2008 ACS data. The results are shown in more detail in Appendix Table D4.1 and Appendix Table D4.2.
28. See Blau (1998).

5. Public Attitudes toward Science

1. National Science Board (2010).
2. Cicerone (2009).
3. Jenkins (1999); National Science Board (2010); Smith (2003); J. Miller (2010).
4. Mooney and Kirshenbaum (2009).
5. J. Miller (1998).
6. J. Miller (2010).
7. National Science Board (2010, p. 7-23).
8. J. Miller (2010).
9. Mooney and Kirshenbaum (2009).
10. J. Miller et al. (2006).
11. Kevles (1978); Neal et al. (2008). Also see Chapter 1.
12. Pew Research Center (2009).
13. For a history of science news coverage in the United States, from the development of journalistic science writing in the mid-twentieth century to online coverage and blogs, see Russell (2010).
14. Specifically, we assigned a score of 0 to articles that had no scientific content, 1 to articles that had some scientific content, and 2 to articles that were mainly about science.
15. We obtained the list from Hawes Publications (2010).
16. National Science Board (2010, chapter 7).
17. Pew Research Center (2009).
18. National Science Board (2010, chapter 7).
19. Ibid.
20. Ibid.
21. National Science Board (2010, p. 7-35).
22. Pew Research Center (2009).
23. Ibid.
24. National Science Board (2010, appendix table 7-23).
25. Ibid., p. 7-29.

26. Pew Research Center (2009).
27. National Science Board (2010, p. 7-29).
28. Pion and Lipsey (1981, p. 304).
29. National Science Board (2010, pp. 7-5 and 7-36).
30. OECD (2006).
31. Dawkins (2008); Harris (2006).
32. Pew Research Center (2009, p. 16).
33. Ibid.
34. Ibid., p. 38.
35. Responses to the scientific knowledge questions, by religion, are shown in Appendix Table D5.1.
36. J. Miller et al. (2006).
37. Keeter et al. (2007).
38. Ibid.
39. Pew Research Center (2009, p. 6).
40. Keeter et al. (2007).

6. Does Science Appeal to Students?

1. National Science Board (2010, p. 7-42).
2. National Center for Education Statistics (2009).
3. Ibid.
4. Ibid.
5. National Science Board (2010, p. 1-37).
6. Correll (2001); Jacobs and Eccles (1992).
7. Correll (2001); Mau (2003).
8. Correll (2001).
9. Ibid.; Jacobs and Eccles (1992); Eccles et al. (1993).
10. Mead and Métraux (1957).
11. Ibid.
12. For example, P. Miller et al. (2006).
13. Yager and Yager (1985).
14. Ibid.
15. Barman (1999); Beardslee and O'Dowd (1961); Chambers (1983); Finson (2002); Fort and Varney (1989); Parsons (1997); Rahm and Charbonneau (1997).
16. Barman (1999); Finson (2002); Fort and Varney (1989); Parsons (1997).
17. Losh (2010). Losh's study is concerned with changes between 1983 and 2001 in the images of scientists among American adults.
18. Chambers (1983); Dorkins (1977); Jackson (1992); Maoldomhnaigh and Hunt (1988); D. Newton and Newton (1992); L. Newton and Newton (1998); Schibeci and Sorenson (1983); She (1998); Song et al. (1992); Song and Kim (1999).
19. Forbes (2011).
20. Regarding the underrepresentation of women and minorities in science, see National Science Foundation (1986, 2011).
21. Walton and Cohen (2007).
22. P. Miller et al. (2006).

23. Ibid.
24. For an economic explanation for such effects, see Freeman (1971).
25. Xie and Shauman (2003). The sample probabilities associated with transitioning from expecting a science degree to achieving one, by sex and cohort, are shown in Appendix Table D6.1.
26. Trends in the demographic makeup of these student cohorts are shown in Appendix Table D6.2.
27. College majors coded into fields of science are given in Codebooks C, D, and E, available online at http://www.yuxie.com.
28. For the 1972 cohort, students were asked about attending institutions rather than degree attainment. For this data set, we approximate the set of students who anticipate receiving bachelor's degrees with the set of students who report expecting to attend a four-year college or university or graduate or professional school.
29. The (sample) probabilities of expecting a bachelor's degree, by demographic subgroups, are shown in Appendix Table D6.3. The results of multivariate models predicting expectation of a bachelor's degree are shown in Appendix Table D6.4. The unadjusted probabilities by demographic subgroups of expecting a scientific degree, conditional on expecting a bachelor's degree, are shown in Appendix Table D6.5, and the results for the models predicting expectation of a scientific degree, conditional on expecting a bachelor's degree, are shown in Appendix Table D6.6.
30. Throughout, we refer to results that are significant at the 5 percent level as "significant."
31. In the score model, differences by maternal education are jointly but not individually statistically significant, and they do not follow a monotonic pattern.

7. Attainment of Science Degrees

1. Citro and Kalton (1989); Xie and Shauman (2003).
2. WebCASPAR Integrated Science and Engineering Resource System (2010).
3. These and other degree statistics reported in this chapter were calculated based on our own definition of science and engineering, a definition more restrictive than that used by the National Science Board (2008 and 2010). A major difference is that the National Science Board definition includes social/behavioral sciences. See Codebook B (available online at http://www.yuxie.com) for our definition of fields that are considered science and engineering. We note that the National Science Board used different field definitions between the 2008 and 2010 releases of *Science and Engineering Indicators* (National Science Board 2008, 2010). The two editions differ in their estimates of doctoral degrees because they relied on different data sources. The statistics reported in the 2008 edition on doctoral degrees were based on the Survey of Earned Doctorates, whereas those in the 2010 edition were based on a survey of educational institutions, the Completions Survey. We found that the main discrepancy between the two series of estimates lies in an ambiguous category of "other life sciences." For the year 2005, for example, *Science and Engineering Indicators 2008* (National Science

Board 2008, appendix table 2-31) reported 29,751 doctoral degrees in science and engineering, 970 of which were in "other life sciences." *Science and Engineering Indicators 2010* (National Science Board 2010, appendix table 2-28), however, reported 34,468 doctoral degrees in science and engineering for the same year, the difference being attributable mainly to an upwardly revised estimate of 4,763 degrees in "other life sciences." This category is peculiar in that it includes an extremely low percentage of temporary residents (4 percent among doctoral degree recipients in 2005) and is overwhelmingly female (73 percent female among doctoral degree recipients in 2005). A report from the National Science Foundation (Hill et al. 2007, appendix 14) shows that the "other life sciences" include speech/language pathology and audiology, nursing, rehabilitation, and health policy, fields that are unrelated to core biological sciences. For these reasons, we excluded "other life sciences" from our calculations, along with "medical sciences," a category from which "other life sciences" was separated after 1994.

4. Annualized rates of growth are calculated based on an exponential growth model with a fixed growth rate. For degree production estimates prior to 1960, see Price (1963, p. 7). We calculated the growth rate in the scientific labor force from table 5.1 and data from the Bureau of Labor Statistics (2010b). In Chapter 8, we will provide a more thorough discussion of the growth rate of the scientific labor force.

5. Growth in degrees in the mathematical sciences is not surprising given that computer science was included as part of the mathematical sciences in this calculation.

6. Calculation was based on estimates provided by Lee and Mather (2008).

7. The federal agency responsible for collecting doctoral degree completion data changed its classification for doctoral degrees in 2008, beginning to separate research degrees from professional degrees. To keep the WebCASPAR (2010) data comparable between 2008 and earlier years, we combined the number of degrees reported under the old "Doctorate Degrees" category with those under the new "Doctorate Degree—Research/Scholarship."

8. See Xie and Shauman (2003) for a more extensive discussion on this topic. Figure 7.2a in this book updates the comparable data reported in Xie and Shauman (2003, table 7.1).

9. WebCASPAR (2010). The reason for the decline in women's representation among recipients of bachelor's degrees in mathematical sciences is unclear. Xie and Shauman (2003, p. 138) suspect that the field composition of what is grouped under mathematical sciences has changed.

10. See, for example, Borjas (2004, 2005); Bound et al. (2009); Freeman (2006); Freeman et al. (2004); Matthews (2008); National Academy of Sciences et al. (2007).

11. Bound et al. (2009); Borjas (2004).

12. Galama and Hosek (2008, p. 87).

13. Borjas (2005); North (1995).

14. Sana (2010) also finds that most of the growth between 1994 and 2006 in the number of U.S. doctorates conferred annually in science and engineering was due to immigration.

15. Authors' supplemental analyses using WebCASPAR (2010).
16. Benderly (2010).
17. See Fischer and Hout (2006).
18. For effects on earnings, see Becker (1964); Willis and Rosen (1979). For effects on social status, see Blau and Duncan (1967); Raftery and Hout (1993); Sewell et al. (1969). For general impact of education, see Fischer and Hout (2006).
19. Boudon (1974); Bourdieu (1977); Bowles and Gintis (1976); Brand and Xie (2010).
20. Merton (1942, p. 270).
21. Xie (1989a, 1989b).
22. McFadden (1974); Train (2003); Coleman (1990); and Xie and Shauman (1997); Bandura (1986).
23. The time limit of observation to eight years after high school graduation is dictated by the structure of the NELS data set, whose cohort was last interviewed in 2000, only eight years after normative high school graduation. For comparability, we require in all cohorts that the bachelor's degree be received within eight years of the calendar year of high school graduation. The full lists of fields of study considered scientific are given in Codebooks C, D, and E (available online at http://www.yuxie.com) for the National Longitudinal Study of 1972, High School and Beyond, and the National Education Longitudinal Study of 1988, respectively.
24. Detailed information on the construction of the samples is provided in Appendix B. Trends in the demographic composition of the cohorts are shown in Appendix Table D7.1. Because the data we analyze were designed to be representative of a given cohort of high school seniors, the majority of our sample did not receive a bachelor's degree and even fewer went on to obtain graduate degrees. While the samples are large enough for a study of trends in the receipt of bachelor's degrees and the choice of scientific majors among those degrees, we do not have enough cases to present similar analyses for master's and doctoral degree recipients.
25. Xie and Shauman (2003); Xie (1996).
26. This result is consistent with earlier research on the same topic by Lowell and Salzman (2007).
27. The (sample) probabilities receiving a bachelor's degree, by demographic subgroups, are shown in Appendix Table D7.2. The unadjusted probabilities by demographic subgroups of receiving a scientific degree, conditional on receiving a bachelor's degree, are shown in Appendix Table D7.4.
28. The results of multivariate models predicting receipt of a bachelor's degree are shown in Appendix Table D7.3. The multivariate results of models predicting receipt of scientific degree given a bachelor's degree are shown in Appendix Table D7.5.
29. Seymour and Hewitt (1997).
30. Ibid., pp. 33–36.
31. For the role of noncognitive skills in the general labor market, see Heckman et al. (2006).
32. Ibid., p. 33.
33. Of course, this is a gross simplification that is involved in facilitating interpretation of data. See discussion of this in Xie and Shauman (2003, chapter 5).

34. In NSRCG, professional degrees include the JD, LLB, MD, and DDS but exclude the MBA and EdD.

35. The full lists of fields of study considered scientific are given in Codebooks F and G (available online at http://www.yuxie.com) for the NES and NSRCG data, respectively.

36. Transition to further education includes graduates who were either enrolled in or had completed further education by the time of survey. Detailed trends by field can be found in Appendix Table D7.6.

37. Detailed transition rates by the full interaction between gender and field of study are shown in Appendix Table D7.7.

38. This pattern was found by Xie (1996) and Xie and Shauman (2003, chapter 5).

39. Results by gender and field interaction are shown in Appendix Table D7.6.

40. For example, see Lowell and Salzman (2007, p. 31).

41. The *Science and Engineering Indicators* statistic was also about completed graduate degrees, not graduate enrollment, but we do not believe that this is the main source of the disagreement.

42. See the original report (National Science Foundation 2006).

43. National Science Foundation (2006, table 1). The report also shows that degree recipients in life science were more likely to shift out of science than to stay in science at the graduate level. We think that the NSF definition of "other life sciences," like their definition of "science and engineering," was too broad. See note 3.

8. Finding Work in Science

1. Benderly (2010).

2. Ibid.

3. Lowell and Salzman (2007, p. 30).

4. National Science Board (2010, p. 3-6).

5. Bureau of Labor Statistics (2010b).

6. This number refers to the number of workers "with an S&E bachelor's degree or higher." Another commonly used indicator is the number of workers "whose highest degree was in S&E." However, government estimates for the latter indicator seem to have shifted over time. For example, *Science and Engineering Indicators 2008* (National Science Board 2008, p. 3-6) reported the estimate to be 15 million in 2006, whereas *Science and Engineering Indicators 2010* (National Science Board 2010, p. 3-10) revised the estimate to be 12.4 million for the same year.

7. The full list of scientific occupations appears in Codebook A. Fields of study considered scientific are given in Codebook B. Both codebooks are available online at http://www.yuxie.com.

8. National Science Board (2010, table 3-25, p. 3-53).

9. This assumes that the utilization rate of foreign science degrees for science occupations is about 50 percent.

10. Bureau of Labor Statistics (2010b).

11. WebCASPAR (2010). The percentage was higher at the doctoral level but similar at the master's level. As we discussed before, the potential scientist population is primarily determined by workers with bachelor's degrees.

12. National Science Board (2008, p. 3-17).
13. We use a linear extrapolation from years 2000–2007 to estimate the proportion of scientists in the employed labor force in 2008, the size of which was obtained from the Bureau of Labor Statistics (2010b).
14. National Science Board (2010, figure 2-5).
15. More detailed information on the analyses performed in this chapter can be found in Appendix C.
16. The list of fields of study and occupations considered scientific are available in Codebooks F (NES) and G (NSRCG), available online at http://www.yuxie.com.
17. An employment history over a longer term after graduation would be better. However, we do not have comparable longitudinal data for new graduates across years.
18. The more detailed data broken down by field are given in Appendix Table D8.1.
19. Lowell and Salzman (2007, p. 30).
20. We note that only 34–40 percent of working female graduates at the bachelor's level worked in science in 2003 and 2006.
21. See National Science Board (2006, p. 3-12).
22. National Science Board (2010, table 3-16, p. 3-41) for year 2006. The rate varied by field, between 3.6 percent for engineering graduates and 9.9 percent for graduates with degrees in life science.
23. National Science Board (2010, chapter 3).
24. From 55 percent in 1973 to 45 percent in 2006 (National Science Board 2008, table 5-11, p. 5-27). However, these statistics include doctorates in social science and health science. The inclusion of social science and health science increases the rate of academic employment a bit, as social scientists and health doctoral scientists have a slightly higher rate of academic employment than the overall rate and constitute a small rather than a large share of all doctoral degree holders. For consistency with the rest of the chapter, we recalculated the rate after excluding social scientists and health scientists, based on data reported by the National Science Foundation (2009a, table 12).
25. This is our own calculation from the 1 percent PUMS of the 1990 census.
26. This is well known in the sociology of science literature. See a discussion of the high inequality of rewards in science in Chapter 3.
27. Calculated from data reported by the National Science Foundation (2009a, table 12).
28. Calculated from data reported by the National Science Foundation (2009a, table 54).
29. Boss and Eckert (2006, p. 4).
30. Most notably, National Research Council (2005). Also see Benderly (2004), Check (2007), and Galama and Hosek (2007).
31. Trends in academic employment were calculated from *Science and Engineering Indicators 2008* (National Science Board 2008, appendix table 5-17). The growth rate in doctoral degree production was calculated using the same method as in Chapter 7, but for the period 1973–2006.
32. Calculated from data reported by National Science Board (2008, appendix table 5-17).
33. National Postdoctoral Association (2010).

34. Committee on Science, Engineering, and Public Policy (2000).
35. G. Davis (2006).
36. National Science Foundation (2005).
37. Committee on Science, Engineering, and Public Policy (2000).
38. Calculated from data reported by the National Science Board (2010, appendix table 2-32). These numbers are noticeably higher (more than double) than alternative estimates also reported by the National Science Board (2008, appendix table 5-17), presumably due to definitional differences. The higher numbers likely included persons that were considered nonfaculty employees by alternative definitions.
39. Calculated from data released by the National Science Foundation (2009b, table 28).
40. National Academy of Sciences et al. (2000).
41. Nerad and Cerny (1999); National Research Council (2005).
42. For the discussion given below, we draw from Benderly (2004, 2010), Committee on Science, Engineering, and Public Policy (2000), National Research Council (2005), National Science Foundation (1998, 2005), and Nerad and Cerny (1999).
43. Examples are plentiful, including Albert Einstein. Also see National Research Council (2005, p. 18).
44. National Research Council (2005, p. 15).
45. Benderly (2010).
46. National Research Council (2005, p. vii).
47. Kreeger (2004).
48. National Heart Lung and Blood Institute et al. (2010).
49. Benderly (2010).
50. Indeed, one of the key recommendations of the National Research Council (2005) report is to ask the NIH to set a five-year limit on the use of funding to support postdoctoral researchers.
51. Zumeta and Raveling (2002).
52. Lowell and Salzman (2007, p. 37).
53. Hoffer et al. (2007); National Science Foundation (1998, 2005).

Conclusion

1. National Academy of Sciences et al. (2007); Atkinson (1990).
2. Galama and Hosek (2007); Lowell and Salzman (2007).
3. For example, it has been argued that immigration of scientists may suppress the earnings of scientists by increasing their supply relative to a fixed demand (Borjas 2005; North 1995).
4. Fischer and Hout (2006); Goldin and Katz (2008); Heckman and Kruger (2003).
5. For a review and a critique of the peer-review system in science, see Chubin and Hackett (1990).
6. National Research Council (2005, p. viii).
7. Note that the above-quoted passage was written jointly by Thomas R. Cech, a Nobel laureate and then president of Howard Hughes Medical Institute, and Bruce Alberts, then president of the National Academy of Sciences.

8. Such concerns are expressed by J. Cole (2009) and the National Academy of Sciences et al. (2007).

9. See Page (2007) for efficiency gains due to intellectual diversity.

10. Most notably, see the Bush (1945) report for the establishment of the National Science Foundation. Also see Kevles (1977, 1978).

11. The three features we list here are similar to the set of four "institutional imperatives" described by Merton (1942): universalism, communism, disinterestedness, and organized skepticism.

12. Lipset (1963); McElroy (1999); Nye (1960); Tocqueville (1904).

13. This is true despite a history of racial conflicts and gender inequality. Blau and Duncan (1967); Hout (1988); Lipset (1963); Xie and Goyette (2003).

14. Lipset (1963, pp. 121–122).

15. Alba and Nee (2003); Zeng and Xie (2004).

16. Fischer and Hout (2006); Glazer (2003); Hodgson (2009).

17. Long and Fox (1995); Long et al. (1979).

18. Merton (1942).

19. Lipset (1963, p. 68).

20. Tocqueville (1904, p. 624).

21. Lipset (1963, pp. 175–176).

22. McElroy (1999, p. 148).

23. See, for example, the classic book by Jencks et al. (1972).

24. McElroy (1999, p. 155).

25. Svallfors (1997).

26. Lipset (1963).

27. The following points are drawn from Lipset (1963) and McElroy (1999).

28. Also see Turner (1960).

29. Lipset (1963, p. 176) argued that Americans even forgive individuals who succeeded through illegal means.

30. A notable example was that Einstein became a popular celebrity when he first visited America. Isaacson (2007).

31. Cole and Cole (1973).

32. Nye (1960).

33. Cooper (1780).

34. Drawing causal conclusions based on observational data always requires unverifiable assumptions. See Morgan and Winship (2007).

35. In the 2007 RAND conference proceedings report (Galama and Hosek, eds.), several economists discussed this recommendation to increase the demand for scientists (Freeman 2007; Stephan 2007; Teitelbaum 2007).

36. Benderly (2010).

37. National Research Council (2005).

38. See J. Cole (2009) for the history of the important role that immigrants have played in American science.

39. This discussion draws from Goldin and Katz (2008).

40. Hout (2010).

41. Fischer and Hout (2006); Goldin and Katz (2008); Heckman and Kruger (2003).

42. Gamoran (2001); Heckman (2006); Kao and Thompson (2003). Note that James Heckman appropriately emphasizes the importance of education in early ages. See Heckman and Kruger (2003).
43. Hout (2010).
44. Arum and Roksa (2011).

Appendix A

1. Respondents with a bachelor's degree are much more common than respondents with an advanced degree, but respondents with seventeen years of education represent a much larger proportion of respondents with an advanced degree than they do respondents with a bachelor's degree. See Jaeger (1997).
2. Bureau of Labor Statistics (2010a).

Appendix B

1. In accordance with NCES policy, unweighted numbers from High School and Beyond (HS&B) and the National Education Longitudinal Study of 1988 (NELS) are rounded to the nearest ten.
2. The time limit of observation to eight years after high school graduation is dictated by the structure of the NELS data set, the cohort that was last interviewed in 2000, only eight years after high school graduation. For comparability, we require in all cohorts that the bachelor's degree be received within eight years of the calendar year of high school graduation.
3. We considered including social science as part of science. However, since most policy discussions about a possible shortfall in scientific labor force have focused almost exclusively on natural science and engineering, we follow the decision in Xie and Shauman (2003) to exclude social science.
4. In HS&B and NELS, family income is also ascertained in an earlier wave of the survey. For these cohorts, the respondent's reported family income in a previous survey wave was also used in the imputation, when available.

References

Abelson, Philip. 1990. "The Need to Improve the Image of Chemistry." *Science* 249:225.

Adams, Jonathan. 2007. "Scientific Wealth and Scientific Investments of Nations." In *Perspectives on U.S. Competitiveness in Science and Technology*, edited by T. Galama and J. Hosek, 37–48. Santa Monica, CA: RAND.

Adams, Jonathan, and David Pendlebury. 2010. *Global Research Report: United States*. Leeds, UK: Thomson Reuters. Retrieved February 22, 2011 (http://researchanalytics.thomsonreuters.com/m/pdfs/globalresearchreport -usa.pdf).

Alba, Richard, and Victor Nee. 2003. *Remaking the American Mainstream: Assimilation and Contemporary Immigration*. Cambridge, MA: Harvard University Press.

Allison, Paul D., and J. Scott Long. 1987. "Interuniversity Mobility of Academic Scientists." *American Sociological Review* 52:643–652.

Arum, Richard, and Josipa Roksa. 2011. *Academically Adrift: Limited Learning on College Campuses*. Chicago: University of Chicago Press.

Atkinson, Richard D. 1990. "Supply and Demand for Scientists and Engineers: A National Crisis in the Making." *Science* 248:425–432.

Atkinson, Richard D. 1996. "The Numbers Game and Graduate Education." Presented at Conference on Graduation Education in the Biological Sciences in the 21st Century, October 2, San Francisco, CA. Retrieved March 9, 2011 (http://www.ucop.edu/ucophome/pres/comments/numbers.html).

Austin, Jim. 2002. "Demanding More Scientists." *Science*, November 29. Retrieved June 11, 2007 (http://sciencecareers.sciencemag.org/career_development/pre vious_issues/articles/2100/demanding_more_scientists/).

Australia Senate. 2006. "Senate Official Hansard." No. 12, October 17. *Commonwealth of Australia, Parliamentary Debates*. Retrieved June 18, 2010 (www.aph .gov.au/hansard/senate/dailys/ds171006.pdf).

Bandura, Albert. 1986. *Social Foundations of Thought and Action: A Social Cognitive Theory*. Englewood Cliffs, NJ: Prentice-Hall.

Barman, Charles R. 1999. "Students' Views about Scientists and School Science: Engaging K–8 Teachers in a National Study." *Journal of Science Teacher Education* 10:43–54.

Bean, Frank D., and Mark A. Leach. 2005. "A Critical Disjuncture? The Culmination of Post–World War II Socio-Demographic and Economic Trends in the United States." *Journal of Population Research* 22:63–78.

Beardslee, David C., and Donald D. O'Dowd. 1961. "The College-Student Image of Scientists." *Science* 133:997–1001.

Becker, Gary S. 1964. *Human Capital: A Theoretical and Empirical Analysis, with Special Reference to Education.* New York: Columbia University Press.

Becker, Gary S. 1981. *A Treatise on the Family.* Cambridge, MA: Harvard University Press.

Bell, Daniel. 1976. *The Coming of Post-Industrial Society: A Venture in Social Forecasting.* New York: Basic Books.

Bellis, Mary. 2010. "History & Bios—Famous Inventions & Famous Inventors." *About.com.* Retrieved August 30, 2010 (http://inventors.about.com/od/famous inventions/u/history_biography.htm).

Ben-David, Joseph. 1971. *The Scientist's Role in Society: A Comparative Study.* Chicago: University of Chicago Press.

Benderly, Beryl Lieff. 2004. "The Incredible Shrinking Tenure Track." *Science Career Magazine.* Retrieved June 20, 2010 (http://sciencecareers.sciencemag.org /career_development/previous_issues/articles/3150/the_incredible_shrinking _tenure_track).

Benderly, Beryl Lieff. 2010. "The Real Science Gap." *Miller-McCune Online.* Retrieved June 14, 2010 (http://www.miller-mccune.com/science/the-real-science -gap-16191/).

Berry, Colin. 1981. "The Nobel Scientists and the Origins of Scientific Achievement." *British Journal of Sociology* 32:381–391.

Biagioli, Mario. 1993. *Galileo Courtier: The Practice of Science in the Culture of Absolutism.* Chicago: University of Chicago Press.

Blanchard, Emily, John Bound, and Sarah Turner. 2009. "Opening (and Closing) Doors: Country-Specific Shocks in U.S. Doctorate Education." Research Report No. 09-674, Population Studies Center, University of Michigan, Ann Arbor.

Blau, Francine D. 1998. "Trends in the Well-Being of American Women, 1970–1995." *Journal of Economic Literature* 36:112–165.

Blau, Peter, and Otis Dudley Duncan, with Andrea Tyree. 1967. *The American Occupational Structure.* New York: Wiley.

Borjas, George J. 2004. "Do Foreign Students Crowd Out Native Students from Graduate Programs?" Working Paper No. 10349, National Bureau of Economic Research, Chicago. Retrieved June 15, 2010 (http://www.nber.org/papers /w10349).

Borjas, George J. 2005. "The Labor-Market Impact of High-Skill Immigration." *American Economic Review* 95 (2): 56–60.

Boskin, Michael J. 1974. "A Conditional Logit Model of Occupational Choice." *Journal of Political Economy* 82:389–398.

Boss, Jeremy M., and Susan H. Eckert. 2006. *Academic Scientists at Work: Navigating the Biomedical Research Career.* Boston: Springer.

Boudon, Raymond. 1974. *Education, Opportunity and Social Inequality: Changing Prospects in Western Society.* New York: John Wiley & Sons.

Bound, John, Sarah Turner, and Patrick Walsh. 2009. "Internationalization of U.S. Doctorate Education." Research Report No. 09-675, Population Studies Center, University of Michigan, Ann Arbor.

Bourdieu, Pierre. 1977. "Cultural Reproduction and Social Reproduction." In *Power and Ideology in Education*, edited by J. Karabel and A. H. Halsey. New York: Oxford University Press.

Bowles, Samuel, and Herbert Gintis. 1976. *Schooling in Capitalist America: Educational Reform and the Contradictions of Economic Life*. New York: Basic Books.

Brand, Jennie E., and Yu Xie. 2010. "Who Benefits Most from College? Evidence for Negative Selection in Heterogeneous Economic Returns to Higher Education." *American Sociological Review* 75:273–302.

Braverman, Harry. 1974. *Labor and Monopoly Capital: The Degradation of Work in the Twentieth Century*. New York: Monthly Review Press.

Bruce, Mary. 2010. "China Debuts at Top of International Education Rankings." *ABC News/Politics*, December 7. Retrieved February 28, 2011 (http://abcnews .go.com/Politics/china-debuts-top-international-education-rankings/story ?id=12336108&page=1).

Bulmer, Michael. 2003. *Francis Galton: Pioneer of Heredity and Biometry*. Baltimore: Johns Hopkins University Press.

Bureau of Labor Statistics. 2010a. *Consumer Price Index*. Washington, DC: United States Department of Labor. Retrieved March 11, 2011 (ftp://ftp.bls.gov/pub /special.requests/cpi/cpiai.txt).

Bureau of Labor Statistics. 2010b. "Labor Force Statistics from the Current Population Survey." Retrieved May 12, 2011 (ftp://ftp.bls.gov/pub/special.requests/lf /aat1.txt).

Bush, Vannevar. 1945. *Science, the Endless Frontier: A Report to the President on a Program for Postwar Scientific Research*. Reprinted 1960, NSF 60-40. Washington DC: National Science Foundation.

Butz, William P., Gabrielle A. Bloom, Mihal E. Gross, Terrence K. Kelly, Aaron Kofner, and Helga E. Rippen. 2003. "Is There a Shortage of Scientists and Engineers? How Would We Know?" RAND Science and Technology Issue Paper 241. Santa Monica, CA: RAND. Retrieved October 24, 2011 (http://www.rand .org/pubs/issue_papers/IP241.html).

Carlson, W. Bernard. 1997. "Innovation and the Modern Corporation: From Heroic Invention to Industrial Science." In *Science in the Twentieth Century*, edited by J. Krige and D. Pestre, 203–226. Amsterdam: Harwood Academic Publishers.

Chambers, David W. 1983. "Stereotypic Images of the Scientist: The Draw-a-Scientist Test." *Science Education* 67:255–265.

Check, Erika. 2007. "More Biologists but Tenure Stays Static." *Nature* 448:848–849. Retrieved June 20, 2010 (http://www.nature.com/nature/journal/v448 /n7156/full/448848a.html).

Chubin, Daryl E., and Edward J. Hackett. 1990. *Peerless Science: Peer Review and U.S. Science Policy*. Albany: State University of New York Press.

Cicerone, Ralph J. 2009. "How Healthy Is Science in the United States?" Speech at the 146th annual meeting of the National Academy of Sciences, April 25–28,

Washington, DC. Retrieved May 9, 2011 (http://www.nasonline.org/site/Doc Server/2009speech.pdf?docID=55601).

Citro, Constance F., and Graham Kalton, eds. 1989. *Surveying the Nation's Scientists and Engineers: A Data System for the 1990s.* Washington, DC: National Academy Press.

Clark, Ronald W. 1983. *Benjamin Franklin: A Biography.* New York: Random House.

Clark, William. 2003. "The Pursuit of the Prosopography of Science." In *Cambridge History of Science,* vol. 4, edited by R. Porter, 211–237. Cambridge: Cambridge University Press.

Cole, Jonathan R. 1979. *Fair Science: Women in the Scientific Community.* New York: Free Press.

Cole, Jonathan R. 2009. *The Great American University: Its Rise to Preeminence, Its Indispensable National Role, Why It Must Be Protected.* New York: PublicAffairs.

Cole, Jonathan, and Stephen Cole. 1973. *Social Stratification in Science.* Chicago: University of Chicago Press.

Cole, Stephen. 2004. "Merton's Contribution to the Sociology of Science." *Social Studies of Science* 34:829–844.

Coleman, James S. 1990. *Foundations of Social Theory.* Cambridge, MA: Belknap Press of Harvard University Press.

Committee on Science, Engineering, and Public Policy. 2000. *Enhancing the Postdoctoral Experience for Scientists and Engineers.* Washington, DC: National Academies Press.

Committee on Underrepresented Groups and the Expansion of the Science and Engineering Workforce Pipeline; Committee on Science, Engineering, and Public Policy; Policy and Global Affairs; National Academy of Sciences; National Academy of Engineering; and Institute of Medicine. 2011. *Expanding Underrepresented Minority Participation: America's Science and Technology Talent at the Crossroads.* Washington, DC: National Academies Press.

Cooper, Samuel. (1780) 2010. "A Sermon on the Commencement of the Constitution, October 25, 1780." TeachingAmericanHistory.org. Retrieved August 28, 2010 (http://teachingamericanhistory.org/library/index.asp?document=598).

Correll, Shelley J. 2001. "Gender and the Career Choice Process: The Role of Biased Self-Assessments." *American Journal of Sociology* 106:1691–1730.

DARPA. 2011. "Our Work." Retrieved July 29, 2011 (http://www.darpa.mil/our _work/).

Davis, G. 2006. "Improving the Postdoctoral Experience: An Empirical Approach." In *Science and Engineering Careers in the United States: An Analysis of Markets and Employment,* edited by R. Freeman and D. Goroff, 99–130. Chicago: NBER/University of Chicago Press.

Davis, James A. 1964. *Great Aspirations: The Graduate School Plans of America's College Seniors.* Chicago: Aldine.

Davis, James A. 1965. *Undergraduate Career Decisions: Correlates of Occupational Choice.* Chicago: Aldine.

Dawkins, Richard. 2008. *The God Delusion.* New York: First Mariner Books.

Daymont, Thomas N., and Paul J. Andrisani. 1984. "Job Preferences, College Major, and the Gender Gap in Earnings." *Journal of Human Resources* 19:408–428.

Diamond, Jared M. 1999. *Guns, Germs, and Steel: The Fates of Human Societies.* New York: W. W. Norton.

Dillon, Sam. 2010. "Top Test Scores from Shanghai Stun Educators." *New York Times,* December 7. Retrieved February 28, 2011. (http://www.nytimes.com /2010/12/07/education/07education.html?_r=3&src=me&ref=homepage).

DiPrete, Thomas A., and Gregory M. Eirich. 2006. "Cumulative Advantage as a Mechanism for Inequality: A Review of Theoretical and Empirical Developments." *Annual Review of Sociology* 32:271–297.

Dorkins, Huw. 1977. "Sixth Form Attitudes to Science." *New Scientists* 75:523–524.

Duncan, Otis Dudley. 1994. "Inscribed Thoughts 1993–1994." Retrieved June 18, 2010 (http://personal.psc.isr.umich.edu/~yuxie/FTP/duncan.htm).

Duncan, Otis Dudley, David L. Featherman, and Beverly Duncan. 1972. *Socioeconomic Background and Achievement.* New York: Academic Press.

Eccles, Jacquelynne, Allan Wigfield, Rena D. Harold, and Phyllis Blumenfeld. 1993. "Age and Gender Differences in Children's Self- and Task Perceptions during Elementary School." *Child Development* 64:830–847.

Education for Innovation Initiative. 2008. *Tapping America's Potential.* "Gaining Momentum, Losing Ground." Progress Report. Retrieved November 9, 2011 (http://www.eric.ed.gov/PDFS/ED502334.pdf).

Eiduson, Bernice T. 1962. *Scientists: Their Psychological World.* New York: Basic Books.

Eiduson, Bernice T., and Linda Beckman, eds. 1973. *Science as a Career Choice: Theoretical and Empirical Studies.* New York: Russell Sage Foundation.

Einstein, Albert. 1951. Letter to California student E. Holzapfel. Quoted in Helen Dukas and Banesh Hoffman, *Albert Einstein, the Human Side.* Princeton, NJ: Princeton University Press, 1981.

Ezell, Stephen J., and Robert D. Atkinson. 2008. "RAND's Rose-Colored Glasses: How RAND's Report on U.S. Competitiveness in Science and Technology Gets It Wrong." The Information Technology and Innovation Foundation. Retrieved July 23, 2010 (http://www.itif.org/files/2008-RAND%20Rose-Colored%20 Glasses.pdf).

Featherman, David L., and Robert M. Hauser. 1978. *Opportunity and Change.* New York: Academic Press.

Finson, Kevin D. 2002. "Drawing a Scientist: What We Do and Do Not Know after Fifty Years of Drawings." *School Science and Mathematics* 102:335–345.

Fischer, Claude S., and Michael Hout. 2006. *Century of Difference: How America Changed in the Last One Hundred Years.* New York: Russell Sage Foundation.

Fleischman, Howard L., Paul J. Hopstock, Marisa P. Pelczar, and Brooke E. Shelley. 2010. *Highlights from PISA 2009: Performance of U.S. 15-Year-Old Students in Reading, Mathematics, and Science Literacy in an International Context.* NCES 2011-004. Washington, DC: U.S. Department of Education.

Flynn, James R. 1991. *Asian Americans: Achievement beyond IQ.* Hillsdale, NJ: L. Erlbaum.

Forbes. 2011. "The Richest People In America." Retrieved November 9, 2011 (http://www.forbes.com/wealth/forbes-400/gallery).

Fort, Deborah C., and Heather L. Varney. 1989. "How Students See Scientists: Mostly Male, Mostly White, and Mostly Benevolent." *Science and Children* 26:8–13.

Franklin, Benjamin. (1750) 1914. *Poor Richard's Almanack*. Waterloo, IA: U.S.C. Publishing Company.

Freeman, R. B. 1982. "The Darwin Family." *Biological Journal of the Linnean Society* 17:9–21.

Freeman, Richard B. 1971. *The Market for College-Trained Manpower: A Study in the Economics of Career Choice*. Cambridge, MA: Harvard University Press.

Freeman, Richard B. 2006. "Does Globalization of the Scientific/Engineering Workforce Threaten U.S. Economic Leadership?" In *Innovation Policy and the Economy*, vol. 6, edited by A. B. Jaffe, J. Lerner, and S. Stern, 123–157. Cambridge, MA: MIT Press.

Freeman, Richard B. 2007. "Globalization of the Scientific/Engineering Workforce and National Security." In *Perspectives on U.S. Competitiveness in Science and Technology*, edited by Titus Galama and James Hosek, 81–89. Santa Monica, CA: RAND.

Freeman, Richard B. 2009. "What Does Global Expansion Higher Education Mean for the US?" Working Paper 14962, National Bureau of Economic Research, Chicago. Retrieved August 20, 2010 (http://www.nber.org/papers/w14962).

Freeman, Richard B., Emily Jin, and Chia-Yu Shen, 2004. "Where Do New U.S.-Trained Science-Engineering PhDs Come From?" Working Paper 10554, National Bureau of Economic Research, Chicago. Retrieved June 15, 2010 (http://www.nber.org/papers/w10554).

Friedman, Thomas L. 2005. *The World Is Flat: A Brief History of the Twenty-First Century*. New York: Farrar, Straus, and Giroux.

Galama, Titus, and James R. Hosek, eds. 2007. *Perspectives on U.S. Competitiveness in Science and Technology*. Conference proceedings. Santa Monica, CA: RAND.

Galama, Titus, and James R. Hosek, eds. 2008. *U.S. Competitiveness in Science and Technology*. Santa Monica, CA: RAND.

Galton, Francis. 1869. *Hereditary Genius: An Inquiry into Its Laws and Consequences*. London: Macmillan.

Galton, Francis. 1874. *English Men of Science: Their Nature and Nurture*. London: Macmillan.

Gamoran, Adam. 2001. "American Schooling and Educational Inequality: A Forecast for the 21st Century." *Sociology of Education* 74 (Extra Issue): 135–153.

Garrison, Howard H., Andrea L. Stith, and Susan A. Gerbi. 2005. "Foreign Postdocs: The Changing Face of Biomedical Science in the U.S." *FASEB Journal* 19:1938–1942.

Gascoigne, John. 1995. "The Eighteenth-Century Scientific Community: A Prosopographical Study." *Social Studies of Science* 25:575–581.

Ginzberg, Eli, Sol W. Ginsburg, Sidney Axelrad, and John L. Herma. 1951. *Occupational Choice: An Approach to a General Theory*. New York: Columbia University Press.

Glazer, Nathan. 2003. "On Americans & Inequality." *Daedalus* 132:111–115.

Gleick, James. 2003. *Isaac Newton*. New York: Pantheon Books.

Goldin, Claudia Dale, and Lawrence F. Katz. 2008. *The Race between Education and Technology.* Cambridge, MA: Belknap Press of Harvard University Press.

Gonzales, Patrick, Trevor Williams, Leslie Jocelyn, Stephen Roey, David Kastberg, and Summer Brenwald. 2009. *Highlights from TIMSS 2007: Mathematics and Science Achievement of U.S. Fourth- and Eighth-Grade Students in an International Context* (NCES 2009-001 Revised). Washington, DC: National Center for Education Statistics, Institute for Education Sciences, Department of Education.

Gosling, F. G. 2010. *The Manhattan Project: Making the Atomic Bomb.* National Security History Series. Washington, DC: Department of Energy. Retrieved August 30, 2010 (http://www.energy.gov/media/The_Manhattan_Project_2010.pdf).

Grogan, William R. 1990. "Engineering's Silent Crisis." *Science* 247:381.

Gruber, Howard E. 1981. *Darwin on Man: A Psychological Study of Scientific Creativity.* 2nd ed. Chicago: University of Chicago Press.

Hagstrom, Warren O. 1965. *The Scientific Community.* New York: Basic Books.

Hahn, Roger. 1971. *The Anatomy of a Scientific Institution: The Paris Academy of Sciences, 1666–1803.* Berkeley: University of California Press.

Hall, Marie Boas. 1991. *Promoting Experimental Learning: Experiment and the Royal Society, 1660–1727.* Cambridge: Cambridge University Press.

Harding, Sandra G. 1991. *Whose Science? Whose Knowledge? Thinking from Women's Lives.* Ithaca, NY: Cornell University Press.

Harris, Sam. 2006. *Letter to a Christian Nation.* New York: Alfred A. Knopf.

Heckman, James J. 2006. "Skill Formation and the Economics of Investing in Disadvantaged Children." *Science* 312:1900–1902.

Heckman, James J., and Alan B. Krueger. 2003. *Inequality in America: What Role for Human Capital Policies.* Cambridge, MA: MIT Press.

Heckman, James J., Jora Stixrud, and Sergio Urzua. 2006. "The Effects of Cognitive and Noncognitive Abilities on Labor Market Outcomes and Social Behavior." *Journal of Labor Economics* 24:411–482.

Heston, Alan, Robert Summers, and Bettina Aten. 2009. *Penn World Table Version 6.3.* Philadelphia: Center for International Comparisons of Production, Income and Prices, University of Pennsylvania. Retrieved February 24, 2011 (http://pwt.econ.upenn.edu/php_site/pwt_index.php).

Hill, Derek, Alan I. Rapoport, Rolf F. Lehming, and Robert K. Bell. 2007. *Changing U.S. Output of Scientific Articles: 1988–2003.* NSF 07-320. Arlington, VA: National Science Foundation, Division of Science Resources Statistics. Retrieved June 1, 2010 (http://www.nsf.gov/statistics/nsf07320/content.cfm?pub_id=1878&id=9).

Hilts, Victor L. 1973. "Statistics and Social Science." In *Foundations of Scientific Method: The Nineteenth Century,* edited by R. N. Giere and R. S. Westfall, 206–233. Bloomington: Indiana University Press.

Hilts, Victor L. 1981. *Statist and Statistician: Three Studies in the History of Nineteenth Century English Statistical Thought.* New York: Arno Press.

Hirsch, J. E. 2005. "An Index to Quantify an Individual's Scientific Research Output." *Proceedings of the National Academy of Sciences of the United States of America* 102:16569–16572.

Hirsch, Walter. 1968. *Scientists in American Society.* New York: Random House.

Hodge, Robert W., Paul M. Siegel, and Peter H. Rossi. 1964. "Occupational Prestige in the United States, 1925–63." *American Journal of Sociology* 70:286–302.

Hodgson, Godfrey. 2009. *The Myth of American Exceptionalism.* New Haven, CT: Yale University Press.

Hoffer, T. B., M. Hess, V. Welch Jr., and K. Williams. 2007. *Doctorate Recipients from United States Universities: Summary Report 2006.* Chicago: National Opinion Research Center.

Hollingsworth, J. Rogers, Karl H. Müller, and Ellen Jane Hollingsworth. 2008. "China: The End of the Science Superpowers." *Nature* 454:412–413.

Holt, Charles A., and Susan K. Laury. 2002. "Risk Aversion and Incentive Effects." *American Economic Review* 92:1644–1655.

Hout, Michael 1988. "More Universalism, Less Structural Mobility: The American Occupational Structure in the 1980s." *American Journal of Sociology* 93:1358–1400.

Hout, Michael. 2010. "Rationing Opportunity: The Role of America's Colleges and Universities in Graduation Trends." Presented at the 2010 Annual Meeting of the Population Association of American, April 30, Detroit.

Hughes, Thomas P. 2004. *American Genesis: A Century of Invention and Technological Enthusiasm, 1870–1970.* New York: Penguin.

IMD International. 2005. *World Competitiveness Yearbook.* Lausanne, Switzerland: IMD International.

Isaacson, Walter. 2003. *Benjamin Franklin: An American Life.* New York: Simon & Schuster.

Isaacson, Walter. 2007. *Einstein: His Life and Universe.* New York: Simon & Schuster.

Jackson, T. 1992. "Perceptions of Scientists among Elementary School Children." *Australian Science Teachers Journal* 38:57–61.

Jacobs, Janis E., and Jacquelynne S. Eccles. 1992. "The Impact of Mothers' Gender-Role Stereotypic Beliefs on Mothers' and Children's Ability Perceptions." *Journal of Personality and Social Psychology* 63:932–944.

Jaeger, David A. 1997. "Reconciling the Old and New Census Bureau Education Questions: Recommendations for Researchers." *Journal of Business & Economic Statistics* 15:300–309.

Jencks, Christopher, Marshall Smith, Henry Acland, Mary Jo Bane, David Cohen, Herbert Gintis, Barbara Heyns, and Stephan Michelson. 1972. *Inequality: A Reassessment of the Effect of Family and Schooling in America.* New York: Basic Books.

Jenkins, E. W. 1999. "School Science, Citizenship and the Public Understanding of Science." *International Journal of Science Education* 21:703–710.

Jones, M. Gail, Ann Howe, and Melissa J. Rua. 2000. "Gender Differences in Students' Experiences, Interests, and Attitudes toward Science and Scientists." *Science Education* 84 (2): 180–192.

Kao, Grace, and Jennifer S. Thompson. 2003. "Race and Ethnic Stratification in Educational Achievement and Attainment." *Annual Review of Sociology* 29:417–442.

Katz, Lawrence F., and David H. Autor. 1999. "Changes in the Wage Structure and Earnings Inequality." In *Handbook of Labor Economics,* vol. 3A, edited by O. Ashenfelter and D. Card, 1463–1555. Amsterdam: Elsevier Science, North-Holland.

Kaufman, Senator Edward. 2009. Speech on the Senate Floor, February 27. Washington, DC. Published in *Congressional Record* (http://frwebgate1.access.gpo.gov/cgi-bin/TEXTgate.cgi?WAISdocID=5XHXgk/7/1/0&WAISaction=retrieve).

Keeter, Scott, Gregory Smith, and David Masci. 2007. "Religious Belief and Public Attitudes about Science in the U.S." Pew Research Center. Retrieved August 30, 2010 (http://pewresearch.org/assets/pdf/667.pdf).

Keller, Evelyn Fox. 1996. *Reflections on Gender and Science.* Tenth anniversary paperback ed. New Haven, CT: Yale University Press.

Kevles, Daniel J. 1977. "The National Science Foundation and the Debate over Postwar Research Policy, 1942–1945: A Political Interpretation of *Science—The Endless Frontier.*" *Isis* 68:4–26.

Kevles, Daniel J. 1978. *The Physicists: the History of a Scientific Community in Modern America.* New York: Knopf.

Konrad, Alison M., J. Edgar Ritchie Jr., Pamela Lieb, and Elizabeth Corrigall. 2000. "Sex Differences and Similarities in Job Attribute Preferences: A Meta-Analysis." *Psychological Bulletin* 126:593–641.

Koyré, Alexandre. 1965. *Newtonian Studies.* Cambridge, MA: Harvard University Press.

Kreeger, Karen. 2004. "Postdocs: Healthy Limits." *Nature* 427:178–179. Retrieved August 31, 2010 (http://www.nature.com/nature/journal/v427/n6970/full/nj6970-178a.html).

Kuhn, Thomas S. 1992. *The Copernican Revolution: Planetary Astronomy in the Development of Western Thought.* New ed. Cambridge, MA: Harvard University Press.

Kuhn, Thomas, S. 1996. *The Structure of Scientific Revolutions.* 3rd ed. Chicago: University of Chicago Press.

Lawler, Andrew. 2005. "Summers's Comments Draw Attention to Gender, Racial Gaps." *Science* 307 (5709): 492–493.

Lee, Jennifer, and Frank D. Bean. 2004. "America's Changing Color Lines: Immigration, Race/Ethnicity, and Multiracial Identification." *Annual Review of Sociology* 30:221–242.

Lee, Marlene A., and Mark Mather. 2008. "U.S. Labor Force Trends." *Population Bulletin. Population Reference Bureau.* Retrieved July 28, 2010 (http://www.prb.org/Publications/PopulationBulletins/2008/uslaborforce.aspx).

Lipset, Seymour Martin. 1963. *The First New Nation.* New York: Basic Books.

Livingstone, David N. 2003. *Putting Science in Its Place: Geographies of Scientific Knowledge.* Chicago: University of Chicago Press.

Long, J. Scott, Paul D. Allison, and Robert McGinnis. 1979. "Entrance into the Academic Career." *American Sociological Review* 44:816–830.

Long, J. Scott, and Mary F. Fox. 1995. "Scientific Careers: Universalism and Particularism." *Annual Review of Sociology* 21:45–71.

Losh, Susan Carol. 2010. "Stereotypes about Scientists over Time among U.S. Adults: 1983 and 2001." *Public Understanding of Science* 19 (3): 372–382.

Lowell, B. Lindsay, and Hal Salzman. 2007. "Into the Eye of the Storm: Assessing the Evidence on Science and Engineering Education, Quality, and Workforce Demand." Washington, DC: Urban Institute. Unpublished manuscript.

MacKenzie, Donald A. 1981. *Statistics in Britain, 1865–1930: The Social Construction of Scientific Knowledge.* Edinburgh: Edinburgh University Press.

Mankiw, N. Gregory. 2003. *Macroeconomics.* 5th ed. New York: Worth Publishers.

Manski, Charles F., and David A. Wise. 1983. *College Choice in America.* Cambridge, MA: Harvard University Press.

Maoldomhnaigh, M. C., and A. Hunt. 1988. "Some Factors Affecting the Image of the Scientists Drawn by Older Primary School Pupils." *Research in Science and Technology Education* 6:159–166.

Mare, Robert D. 1980. "Social Background and School Continuation Decisions." *Journal of the American Statistical Association* 75:295–305.

Matthews, Christine M. 2008. *Foreign Science and Engineering Presence in U.S. Institutions and the Labor Force.* Issue brief. Washington, DC: Congressional Research Service.

Mau, Wei-Cheng. 2003. "Factors That Influence Persistence in Science and Engineering Career Aspirations." *Career Development Quarterly* 51:234–243.

McElroy, John Harmon. 1999. *American Beliefs: What Keeps a Big Country and a Diverse People United.* Chicago: Ivan R. Dee.

McFadden, Daniel. 1974. "Conditional Logit Analysis of Qualitative Choice Behavior." In *Frontiers in Econometrics,* edited by P. Zarembka, 105–142. New York: Academic Press.

Mead, Margaret, and Rhoda Métraux. 1957. "Image of the Scientist among High-School Students." *Science* 126 (3270): 384–390.

Merton, Robert K. 1968. "The Matthew Effect in Science: The Reward and Communication System of Science." *Science* 159 (3810): 56–63.

Merton, Robert King. 1942. "The Normative Structure of Science." Reprinted in *The Sociology of Science: Theoretical and Empirical Investigations,* 267–278. Chicago: University of Chicago Press, 1973.

Merton, Robert King. 1970. *Science, Technology & Society in Seventeenth Century England.* New York: Harper & Row.

Mertz, John Theodore. 1965. *History of European Scientific Thought in the Nineteenth Century.* New York: Dover.

Miller, Jon D. 1998. "The Measure of Civic Scientific Literacy." *Public Understanding of Science* 7:203–223.

Miller, Jon D. 2010. "Civic Scientific Literacy: The Role of the Media in the Electronic Era." In *Science and the Media,* edited by D. Kennedy and G. Overholser, 44–63. Cambridge, MA: American Academy of Arts and Sciences.

Miller, Jon D., Eugenie C. Scott, and Shinji Okamoto. 2006. "Public Acceptance of Evolution." *Science* 313 (5788): 765–766.

Miller, Patricia H., Jennifer S. Blessing, and Stephanie Schwartz. 2006. "Gender Differences in High-School Students' Views about Science." *International Journal of Science Education* 28:363–381.

Mooney, Chris, and Sheril Kirshenbaum. 2009. *Unscientific America: How Scientific Illiteracy Threatens Our Future.* New York: Basic Books.

Morgan, Stephen L., and Christopher Winship. 2007. *Counterfactuals and Causal Inference: Methods and Principles for Social Research.* New York: Cambridge University Press.

Mumford, Michael D., Mary Shane Connelly, Ginamarie Scott, Jazmine Espejo, Laura M. Sohl, Samuel T. Hunter, and Katrina E. Bedell. 2005. "Career Experiences and Scientific Performance: A Study of Social, Physical, Life, and Health Sciences." *Creativity Research Journal* 17:105–129.

National Academy of Sciences, National Academy of Engineering, and Institute of Medicine. 2000. *Enhancing the Postdoctoral Experience for Scientists and Engineers: A Guide for Postdoctoral Scholars, Advisers, Institutions, Funding Organizations, and Disciplinary Societies.* Washington, DC: National Academies Press.

National Academy of Sciences, National Academy of Engineering, and Institute of Medicine. 2007. *Rising above the Gathering Storm: Energizing and Employing America for a Brighter Economic Future.* Washington, DC: National Academies Press.

National Academy of Sciences, National Academy of Engineering, and Institute of Medicine. 2010. *Rising above the Gathering Storm: Rapidly Approaching Category 5.* Washington, DC: National Academies Press.

National Aeronautics and Space Administration, Historical Staff. 1963. *The Impact of Sputnik I: Case-Study of American Public Opinion at the Break of the Space Age, October 4, 1957.* NASA Historical Note No. 22. Washington, DC.

National Center for Education Statistics, U.S. Department of Education. 2009. *The Nation's Report Card: NAEP Trends in Academic Progress.* NCES 2009-479. Washington, DC: Institute of Education Sciences. Retrieved August 15, 2011 (http://nces.ed.gov/nationsreportcard/pubs/main2008/2009479.asp).

National Heart Lung and Blood Institute, National Institutes of Health, U.S. Department of Health and Human Services. 2010. "FY 2010 Funding and Operating Guidelines: National Research Service Awards and Career Development Awards." Retrieved July 26, 2010 (http://www.nhlbi.nih.gov/funding/policies/nrsa.htm).

National Institutes of Health. 2010. *The NIH Almanac—Appropriations.* Bethesda, MD. Retrieved August 30, 2010 (http://www.nih.gov/about/almanac/appropriations/index.htm).

National Postdoctoral Association. 2010. "What Is a Postdoc?" Washington, DC. Retrieved June 23, 2010 (http://www.nationalpostdoc.org/policy/what-is-a-postdoc).

National Research Council. 2005. *Bridges to Independence: Fostering the Independence of New Investigators in Biomedical Research.* Washington, DC: National Academies Press.

National Science Board. 2006. *Science and Engineering Indicators 2006.* Arlington, VA: National Science Foundation. Retrieved June 15, 2010 (http://www.nsf.gov/statistics/seind06/).

National Science Board. 2008. *Science and Engineering Indicators 2008.* 2 vols. NSB 08-01 and NSB 08-01A. Arlington, VA: National Science Foundation. Retrieved June 15, 2010 (http://www.nsf.gov/statistics/seind08/).

National Science Board. 2010. *Science and Engineering Indicators 2010.* Arlington, VA: National Science Foundation (NSB 10-01). Retrieved June 15, 2010 (http://www.nsf.gov/statistics/seind10/).

National Science Foundation. 1986. *Women and Minorities in Science and Engineering.* NSF Working Paper 86-301. Arlington, VA: National Science Foundation.

National Science Foundation. 1998. *Survey of Earned Doctorates Report.* Arlington, VA: National Science Foundation.

National Science Foundation. 2005. *Survey of Earned Doctorates Report.* Arlington, VA: National Science Foundation.

National Science Foundation. 2006. *What Do People Do after Earning a Science and Engineering Bachelor's Degree?* NSF Working Paper 06-324. Arlington, VA: National Science Foundation. Retrieved June 15, 2010 (http://www.nsf.gov/statistics /infbrief/nsf06324/).

National Science Foundation. 2009a. *Characteristics of Doctoral Scientists and Engineers in the United States: 2006. Detailed Statistical Tables.* NSF 09-317. Arlington, VA: National Science Foundation. Retrieved June 20, 2010 (http://www.nsf .gov/statistics/nsf09317/).

National Science Foundation. 2009b. *Doctorate Recipients from U.S. Universities: Summary Report 2007–08.* Special Report NSF 10-309. Arlington, VA: National Science Foundation. Retrieved June 15, 2010 (http://www.nsf.gov/statistics /nsf10309/).

National Science Foundation. 2010. *FY 2011 Budget Request to Congress.* Arlington, VA: National Science Foundation. Retrieved August 20, 2010 (http://www.nsf .gov/about/budget/fy2011/table.jsp#quan).

National Science Foundation. 2011. *Women, Minorities, and Persons with Disabilities in Science and Engineering.* NSF Working Paper 11-309. Arlington, VA: National Science Foundation.

Neal, Homer A., Tobin L. Smith, and Jennifer B. McCormick. 2008. *Beyond Sputnik: U.S. Science Policy in the Twenty-First Century.* Ann Arbor: University of Michigan Press.

Nerad, Maresi, and Joseph Cerny. 1999. "Postdoctoral Patterns, Career Advancement, and Problems." *Science,* September 3, 1533–1535. Retrieved June 20, 2010 (http:// www.sciencemag.org/cgi/content/full/285/5433/1533).

Newton, Douglas P., and Lynn D. Newton. 1992. "Young Children's Perceptions of Science and the Scientist." *International Journal of Science Education* 14:331–348.

Newton, Isaac. 1676. Letter to Robert Hooke. February 5. Quoted in *Wikipedia.* 2011. "Standing on the Shoulders of Giants." Retrieved November 8, 2011 (http:// en.wikipedia.org/wiki/Standing_on_the_shoulders_of_giants).

Newton, Isaac. (1687) 1999. *The Principia: Mathematical Principles of Natural Philosophy.* Translated by I. Bernard Cohen and Anne Whitman. Berkeley: University of California Press.

Newton, Lynn D., and Douglas P. Newton. 1998. "Primary Children's Conceptions of Science and the Scientist: Is the Impact of a National Curriculum Breaking down the Stereotype?" *International Journal of Science Education* 20:1137–1149.

New York Times Best Seller List. 1950–2007. Hawes Publications. Retrieved June 10, 2010 (http://www.hawes.com).

Nisbett, Richard E. 2009. *Intelligence and How to Get It.* New York: W. W. Norton.

Nobelprize.org. Retrieved September 7, 2010 (http://nobelprize.org).

Noble, David F. 1977. *America by Design: Science, Technology, and the Rise of Corporate Capitalism.* New York: Knopf.

North, David S. 1995. *Soothing the Establishment: The Impact of Foreign-Born Scientists and Engineers on America.* Lanham, MD: University Press of America.

Nye, Russel Blaine. 1960. *The Cultural Life of the New Nation: 1776–1830.* New York: Harper & Row.

Obama, Barack. 2009. "Remarks by the President at the National Academy of Sciences Annual Meeting." April 27. Washington, DC. Retrieved May 25, 2011 (http://www.whitehouse.gov/the_press_office/Remarks-by-the-President-at -the-National-Academy-of-Sciences-Annual-Meeting/).

Obama, Barack. 2011. "State of the Union 2011: Winning the Future." Retrieved May 25, 2011 (http://www.whitehouse.gov/state-of-the-union-2011).

OECD [Organisation for Economic Co-Operation and Development], Programme for International Student Assessment. 2007. *PISA 2006: Science Competencies for Tomorrow's World,* vol. 1. Paris: OECD Publishing.

OECD [Organisation for Economic Co-Operation and Development], Programme for International Student Assessment. 2010. *PISA 2009: What Students Know and Can Do: Student Performance in Reading, Mathematics, and Science,* vol. 1. Paris: OECD Publishing.

OECD [Organisation for Economic Co-Operation and Development], UNESCO Institute for Statistics. 2005. *Education Trends in Perspective: Analysis of the World Education Indicators.* Paris: OECD Publishing.

Oxford English Dictionary. 2010. "Science." Retrieved October 24, 2011 (http://www .oed.com/view/Entry/172672?redirectedFrom=science#eid).

Oyer, Paul. 2007. "Some Thoughts on the 'Gathering Storm,' National Security, and the Global Market for Scientific Talent." In *Perspectives on U.S. Competitiveness in Science and Technology,* edited by T. Galama and J. Hosek, 113–119. Santa Monica, CA: RAND.

Page, Scott E. 2007. *The Difference: How the Power of Diversity Creates Better Groups, Firms, Schools, and Societies.* Princeton, NJ: Princeton University Press.

Parsons, Eileen Carlton. 1997. "Black High School Females' Images of the Scientist: Expression of Culture." *Journal of Research in Science Teaching* 34 (7): 745–768.

Pew Research Center. 2009. Press release. "Scientific Achievements Less Prominent Than a Decade Ago: Public Praises Science; Scientists Fault Public, Media." Washington, DC: Pew Research Center. Retrieved March 17, 2010 (http://people -press.org/reports/pdf/528.pdf).

Pindyck, Robert S., and Daniel L. Rubinfeld. 2005. *Microeconomics.* 6th ed. Upper Saddle River, NJ: Pearson Prentice Hall.

Pion, Georgine M., and Mark W. Lipsey. 1981. "Public Attitudes toward Science and Technology: What Have the Surveys Told Us?" *Public Opinion Quarterly* 45:303–316.

Price, Derek J. de Solla. 1963. *Little Science, Big Science.* New York: Columbia University Press.

Pryor, John H., Sylvia Hurtado, Victor B. Saenz, José Luis Santos, and William S. Korn. 2007. *The American Freshman: Forty Year Trends.* Los Angeles: Higher Education Research Institute, University of California.

Raftery, Adrian, and Michael Hout. 1993. "Maximally Maintained Inequality: Expansion, Reform, and Opportunity in Irish Education, 1921–75." *Sociology of Education* 66:41–62.

Rahm, Irène, and Paul Charbonneau. 1997. "Probing Stereotypes through Students' Drawings of Scientists." *American Journal of Physics* 65:774–778.

Reiss, Albert J., Jr. 1961. *Occupations and Social Status.* New York: Free Press of Glencoe.

Research!America. 2007. "Americans Support Bridging the Sciences." Retrieved July 23, 2010 (http://www.researchamerica.org/uploads/btspollreport.pdf).

Rever, Philip R. 1973. *Scientific and Technical Careers: Factors Influencing Development during the Educational Years.* Iowa City: American College Testing Program.

Richards, Robert J. 1987. *Darwin and the Emergence of Evolutionary Theories of Mind and Behavior.* Chicago: University of Chicago Press.

Roche, Daniel. 1978. *Le Siècle des Lumières en Province. Académies et Académiciens Provinciaux, 1689–1789.* Paris: Mouton.

Roe, Anne. 1953. *The Making of a Scientist.* New York: Dodd, Mead.

Romer, Paul M. 1990. "Endogenous Technological Change." *Journal of Political Economy* 98:S71–S102.

Rooney, David, Greg E. Hearn, and Abraham Ninan, eds. 2005. *Handbook on the Knowledge Economy.* Cheltenham, UK: Edward Elgar.

Russell, Cristine. 2010. "Covering Controversial Science: Improving Reporting on Science and Public Policy." In *Science and the Media,* edited by D. Kennedy and G. Overholser, 13–43. Cambridge, MA: American Academy of Arts and Sciences.

Sana, Mariano. 2010. "Immigrants and Natives in U.S. Science and Engineering Occupations, 1994–2006." *Demography* 47:801–820.

Schibeci, Renato A., and Irene Sorenson. 1983. "Elementary School Children's Perceptions of Scientists." *School Science and Mathematics* 83:14–20.

Sewell, William H., Archibald O. Haller, and Alejandro Portes. 1969. "The Educational and Early Occupational Attainment Process." *American Sociological Review* 34:82–92.

Sewell, William H., and Robert M. Hauser. 1975. *Education, Occupation, and Earnings: Achievement in the Early Career.* New York: Academic Press.

Seymour, Elaine, and Nancy M. Hewitt. 1997. *Talking about Leaving: Why Undergraduates Leave the Sciences.* Boulder, CO: Westview Press.

Shapin, Steven. 1994. *A Social History of Truth: Civility and Science in Seventeenth-Century England.* Chicago: University of Chicago Press.

Shapin, Steven. 1996. *The Scientific Revolution.* Chicago: University of Chicago Press.

Shapin, Steven. 2008. *The Scientific Life: A Moral History of a Late Modern Vocation.* Chicago: University of Chicago Press.

Shapin, Steven, and Simon Schaffer. 1985. *Leviathan and the Air-Pump: Hobbes, Boyle, and the Experimental Life.* Princeton, NJ: Princeton University Press.

She, Hsiao-ching. 1998. "Gender and Grade Level Differences in Taiwan Students' Stereotypes of Science and Scientists." *Research in Science & Technological Education* 16:125–135.

Smith, Howard A. 2003. "Public Attitudes towards Space Science." *Space Science Reviews* 105:493–505.

Smith, Tom W., Peter Marsden, Michael Hout, and Jibum Kim. 2011. *General Social Surveys, 1972–2010* [machine-readable data file]. Chicago: National Opinion Research Center.

Solow, Robert M. 1957. "Technical Change and the Aggregate Production Function." *Review of Economics and Statistics* 39:312–320.

Somsen, Geert J. 2008. "A History of Universalism: Conceptions of the Internationality of Science from the Enlightenment to the Cold War." *Minerva* 46:361–379.

Song, J., S. J. Pak, and K. A. Jang. 1992. "Attitudes of Boys and Girls in Elementary and Secondary Schools towards Science Lessons and Scientists." *Journal of the Korean Association for Research in Science Education* 12:109–118.

Song, Jinwoong, and Kwang-Suk Kim. 1999. "How Korean Students See Scientists: The Images of the Scientist." *International Journal of Science Education* 21:57–77.

Spain, Daphne, and Suzanne M. Bianchi. 1996. *Balancing Act: Motherhood, Marriage, and Employment among American Women.* New York: Russell Sage Foundation.

Starr, Paul. 1982. *The Social Transformation of American Medicine.* New York: Basic Books.

Stephan, Paula E. 2007. "Comments on the 'Gathering Storm' and Its Implications for National Security." In *Perspectives on U.S. Competitiveness in Science and Technology,* edited by T. Galama and J. Hosek, 101–106. Santa Monica, CA: RAND.

Stevens, Gillian, and David L. Featherman. 1981. "A Revised Socioeconomic Index of Occupational Status." *Social Science Research* 10:364–395.

Stevenson, Harold W., and James W. Stigler. 1992. *The Learning Gap: Why Our Schools Are Failing and What We Can Learn from Japanese and Chinese Education.* New York: Summit Books.

Svallfors, Stefan. 1997. "Worlds of Welfare and Attitudes to Redistribution: A Comparison of Eight Western Nations." *European Sociological Review* 13:283–304.

Teitelbaum, Michael S. 2002. "The U.S. Science and Engineering Workforce: An Unconventional Portrait." Presented at the Government-University-Industry Research Roundtable (GUIRR) Summit, November 12, Washington, DC.

Teitelbaum, Michael S. 2007. "The Gathering Storm and Its Implications for National Security." In *Perspectives on U.S. Competitiveness in Science and Technology,* edited by T. Galama and J. Hosek, 91–100. Santa Monica, CA: RAND.

Tocqueville, Alexis de. 1904. *Democracy in America.* 2 vols. Translated by Henry Reeve. New York: Appleton.

Train, Kenneth. 2003. *Discrete Choice Methods with Simulation.* Cambridge: Cambridge University Press.

Turner, Ralph H. 1960. "Sponsored and Contest Mobility and the School System." *American Sociological Review* 25:855–867.

Twain, Mark. (1889) 1996. *A Connecticut Yankee in King Arthur's Court.* New York: Oxford University Press.

U.S. Census Bureau. 2008. "Historical Statistics of the United States, Colonial Times to 1970." Retrieved May 22, 2008 (http://www.census.gov/compendia/statab/past_years.html).

Vinkler, Peter. 2007. "Eminence of Scientists in the Light of the h-index and Other Scientometric Indicators." *Journal of Information Science* 33:481–491.

Visher, Stephen Sargent. 1947. *Scientists Starred, 1903–1943 in "American Men of Science."* Baltimore: Johns Hopkins University Press.

Waite, Linda J. 1995. "Does Marriage Matter?" *Demography* 32:483–507.

Walton, Gregory M., and Geoffrey L. Cohen. 2007. "A Question of Belonging: Race, Social Fit, and Achievement." *Journal of Personality and Social Psychology* 92:82–96.

Wang, Zuoyue. 2008. *In Sputnik's Shadow: The President's Science Advisory Committee and Cold War America.* New Brunswick, NJ: Rutgers University Press.

Warsh, David. 2006. *Knowledge and the Wealth of Nations: A Story of Economic Discovery.* New York: W. W. Norton.

WebCASPAR Integrated Science and Engineering Resource Data System. 2010. *NSF Survey of Earned Doctorates/Doctorate Records File.* Retrieved June 25, 2010 (https://webcaspar.nsf.gov/).

Weber, Max. (1905) 1958. *The Protestant Ethic and the "Spirit" of Capitalism.* Translated by Talcott Parsons. New York: Charles Scribner's Sons.

Weeden, Kim A., and David B. Grusky. 2005. "The Case for a New Class Map." *American Journal of Sociology* 111:141–212.

West, S. Stewart. 1960. "Sibling Configurations of Scientists." *American Journal of Sociology* 66:268–274.

West, S. Stewart. 1961. "Class Origin of Scientists." *Sociometry* 24:251–269.

Wikipedia. 2009. "Royal Society." Retrieved September 4, 2009 (http://en.wikipedia.org/wiki/Royal_Society).

Wikipedia. 2010. "List of Nobel Laureates by Country." Retrieved September 7, 2010 (http://en.wikipedia.org/wiki/List_of_Nobel_laureates_by_country).

Wikipedia. 2011. "Darwin-Wedgwood Family." Retrieved September 23, 2011 (http://en.wikipedia.org/wiki/Darwin%E2%80%93Wedgwood_family).

Willis, Robert J., and Sherwin Rosen. 1979. "Part 2: Education and Income Distribution." *The Journal of Political Economy* 87 (5): S7–S36.

Xie, Yu. 1989a. "The Process of Becoming a Scientist." PhD dissertation, Department of Sociology, University of Wisconsin, Madison.

Xie, Yu. 1989b. "Structural Equation Models for Ordinal Variables: An Analysis of Occupational Destination." *Sociological Methods & Research* 17:325–352.

Xie, Yu. 1996. "A Demographic Approach to Studying the Process of Becoming a Scientist/Engineer." In *Careers in Science and Technology: An International Perspective,* edited by National Research Council, 43–57. Washington, DC: National Academies Press.

Xie, Yu, and Kimberly Goyette. 2003. "Social Mobility and Educational Choices of Asian Americans." *Social Science Research* 32:467–498.

Xie, Yu, and Kimberly Goyette. 2004. *A Demographic Portrait of Asian Americans.* New York: Russell Sage Foundation and Population Reference Bureau.

Xie, Yu, and Kimberlee A. Shauman. 1997. "Modeling the Sex-Typing of Occupational Choice: Influences of Occupational Structure." *Sociological Methods & Research* 26:233–261.

Xie, Yu, and Kimberlee A. Shauman. 2003. *Women in Science: Career Processes and Outcomes.* Cambridge, MA: Harvard University Press.

Yager, Robert E., and Stuart O. Yager. 1985. "Changes in Perceptions of Science for Third, Seventh, and Eleventh Grade Students." *Journal of Research in Science Teaching* 22:347–358.

Yuasa, Mitsutomo. 1962. "The Shifting Center of Scientific Activity in the West." *Japanese Studies in the History of Science* 1:57–75.

Zeng, Zhen, and Yu Xie. 2004. "Asian-Americans' Earnings Disadvantage Reexamined: The Role of Place of Education." *American Journal of Sociology* 109:1075–1108.

Zuckerman, Harriet. 1977. *Scientific Elite: Nobel Laureates in the United States.* New York: Free Press.

Zumeta, William, and Joyce S. Raveling. 2002. *The Best and Brightest for Science: Is There a Problem Here?* Washington, DC: Commission on Professionals in Science and Technology.

Index